Physics in the Arts
Revised Edition

WHAT IS THE COMPLEMENTARY SCIENCE SERIES?

We hope you enjoy this book. If you would like to read other quality science books with a similar orientation see the order form and reproductions of the front and back covers of other books in the series at the end of this book.

The **Complementary Science Series** is an introductory, interdisciplinary, and relatively inexpensive series of paperbacks for science enthusiasts. The series covers core subjects in chemistry, physics, and biological sciences but often from an interdisciplinary perspective. They are deliberately unburdened by excessive pedagogy, which is distracting to many readers, and avoid the often plodding treatment in many textbooks.

These titles cover topics that are particularly appropriate for self-study although they are often used as complementary texts to supplement standard discussion in textbooks. Many are available as examination copies to professors teaching appropriate courses.

The series was conceived to fill the gaps in the literature between conventional textbooks and monographs by providing real science at an accessible level, with minimal prerequisites so that students at all stages can have expert insight into important and foundational aspects of current scientific thinking.

Many of these titles have strong interdisciplinary appeal and all have a place on the bookshelves of literate laypersons.

Physics in the Arts
Revised Edition

P.U.P.A. Gilbert
University of Wisconsin-Madison

and

W. Haeberli
University of Wisconsin-Madison

AMSTERDAM • BOSTON • HEIDELBERG • LONDON
NEW YORK • OXFORD • PARIS • SAN DIEGO
SAN FRANCISCO • SINGAPORE • SYDNEY • TOKYO

Academic Press is an imprint of Elsevier

The cover image, by Pupa Gilbert, captures all the main themes of this book: light, color, color generating mechanisms, photography, music, sound and waves. White light illuminating the surface of a compact disc is separated into the colors of the spectrum, and photographed. Music is represented by the CD itself, a common music medium, and a sound wave. This oscillation is computer-generated, and similar to a sound wave detected by a microphone and displayed on the screen of an oscilloscope.

Academic Press is an imprint of Elsevier
225 Wyman Street, Waltham, MA 02451, USA
Radarweg 29, PO Box 211, 1000 AE Amsterdam, The Netherlands

Notices
Knowledge and best practice in this field are constantly changing. As new research and experience broaden our understanding, changes in research methods, professional practices, or medical treatment may become necessary.

Practitioners and researchers must always rely on their own experience and knowledge in evaluating and using any information, methods, compounds, or experiments described herein. In using such information or methods they should be mindful of their own safety and the safety of others, including parties for whom they have a professional responsibility.

To the fullest extent of the law, neither the Publisher nor the authors, contributors, or editors, assume any liability for any injury and/or damage to persons or property as a matter of products liability, negligence or otherwise, or from any use or operation of any methods, products, instructions, or ideas contained in the material herein.

Library of Congress Cataloging-in-Publication Data
Gilbert, Pupa.
Physics in the arts / P.U.P.A. Gilbert and W. Haeberli.
    p. cm.
Includes bibliographical references and index.
ISBN 978-0-12-391878-9
1. Light. 2. Music–Acoustics and physics. I. Haeberli, W. (Willy) II. Title.
QC355.3.G55 2011
700.1'05–dc23
                        2011015964

British Library Cataloguing-in-Publication Data
A catalogue record for this book is available from the British Library.

For information on all Academic Press publications
visit our Web site at: *www.elsevierdirect.com*

Printed in China
11 12 13 14 15 9 8 7 6 5 4 3 2 1

Physics in the Arts

*P.U.P.A. Gilbert and
Willy Haeberli*

*University of
Wisconsin-Madison*

Contents

Introduction

Light and Sound

Through light and sound we perceive the world. We have perfected several ways of communicating with each other, and most of these use either light or sound, or both. The highest form of communication is art, and most of the arts use light and/or sound as their expression medium. The visual arts use light, so we can see the works of art; music uses sound so we can hear it; and ballet, movies, and music videos with computer graphics use both.

A deep yet accessible analysis of the physics of light and sound, and how our eyes and ears detect them, is not only intellectually enjoyable, but also useful to understand and interpret the world in which we live, all the phenomena that take place around us, and how we perceive them—in short, how we interface with our planet, its inhabitants, and their creations. Understanding the physics of light and sound may also increase the appreciation for works of art and stimulate the artists among the readers to deepen their knowledge of their media, of how people interface with them, and perhaps inspire new ideas.

Deriving its name from *phusis*, the Greek for "nature," physics is the science that studies naturally occurring phenomena concerning energy or matter. If we can understand a natural phenomenon, then we have contributed to mankind's general knowledge, but often we can also harness it and use it to develop a better life for ourselves and all others on this planet. We think this exciting concept can be exported to the arts and stimulate art productions through the understanding of the physical phenomena underlying them. This very thought motivates our desire to teach physics to artists and humanists and to communicate the physics concepts in simple terms to a more general audience.

Both sound and light are wave phenonema. These are different kinds of waves—mechanical waves in sound and electromagnetic waves in light—but they both oscillate like the surface of a water pond after a stone has been thrown into it. Newton first coined the word *spectrum* of visible light, and then subdivided it into seven colors. Why seven? This choice was arbitrary, but he chose seven by analogy with the seven notes of the musical scale. We now know that the normal human eye can distinguish almost a million colors! For reasons that we will explain later, we will, however, subdivide the visible spectrum into three colors only: red, green, and blue.

Have you ever wondered why the human eye is sensitive to the specific radiation called "light"? Although the sun emits radiation, which, after filtering through the atmosphere, comprises the visible range, there is also a lot of infrared and ultraviolet light that reaches Earth's surface. Why don't we see in those ranges as well? Radiation in the visible range has the most effective energy to interact with the objects of the world. It therefore best informs us on the structure and behavior of objects around us.

The measurable electromagnetic spectrum extends from high-energy gamma rays to low-frequency radio waves. This covers about 30 orders of magnitude in frequency (or energy, or wavelength). From the beginning to the end of what we call visible range (violet to red), the frequency varies by a mere factor of 2. In music, the frequency doubles in one octave. In the measurable electromagnetic spectrum the frequency doubles 100 times ($2^{100} = 10^{30}$). Poetically put, we can only "see one octave" on an idealized electromagnetic piano keyboard of 100 octaves! Yet so much happens in this narrow region because radiation of these energies strongly interacting with electrons generates nonflat, interesting absorption spectra, which in turn generate the million colors to which we are sensitive.

Compared to light, we hear as sound a much larger region of the usable frequencies. The frequency range for audible sound is 20–20,000 oscillations per second or 20–20,000 Hz. Higher frequencies, or ultrasounds, are used for medical imaging (10 MHz) and for communication by dolphins (up to 170 kHz), whales (up to 200 kHz), and bats (up to 120 kHz). Lower frequencies, or infrasounds, are produced by earthquakes, avalanches, volcanoes, nuclear tests, and even elephants (14 Hz)—for communicating with other elephants within a 10 km distance.

The color and sound we perceive do not depend only on the physical, measurable stimuli, but also on the physiological and psychological response of our eyes, ears, and brain to the stimuli. Color and sound, therefore, are best described by psycho-physical parameters. These are, as we will describe in detail later, hue, saturation, and brightness for color, and pitch, loudness, and

The bedroom, 1888 (oil on canvas) by Gogh, Vincent van (1853–90) © Van Gogh Museum, Amsterdam, The Netherlands/The Bridgeman Art Library Nationality/copyright status: Dutch/out of copyright. Ursus Wehrli, Tidying up Van Gogh's Bedroom at Arles. Copyright © Ursus Wehrli. From: Ursus Wehrli, Kunst aufräumen. Copyright © 2002 KEIN & ABER AG, Zürich.

timbre for sound. A trombone and a viola can immediately be distinguished by the listener even if they play the exact same tone. The attribute distinguishing them is called the timbre or, by another analogy with light, "tone color."

The goal of this book is not to tidy up art, rationalize it, and explain it in scientific terms. Other excellent authors did that, as shown in the figures above. Our goal is to add another component—physics—to the enjoyment of art. Understanding the form and function of musical instruments adds to the music. Similarly, understanding color, color vision, and color mixing can only expand the palette of visual artists and the intellectual enjoyment of all people looking at their art.

Both authors collaborated on all parts of the book, however, Pupa Gilbert takes primary responsibility for the part on Light, and Willy Haeberli for that on Sound. We wish you an enjoyable read!

Chapter 1

Light and Light Waves

We all have experience with light and its colors, variations in intensity, reflection, refraction, focusing, and light's characteristic of traveling along a straight line in a homogenous medium. We have, however, no intuitive feeling for the physical nature of light, what it is made of, and how it works. In this chapter we discuss the dual nature of light—how it behaves as both waves and particles—and we will also discuss the electromagnetic spectrum, the speed of light in different media, and polarization.

Until the 1600s everyone thought light was made up of particles. Then, in the 1600s, competing theories of light were proposed by Christiaan Huygens who maintained that light was waves, and Isaac Newton, who insisted it was particles. The debate continued, until finally in the 1900s it was agreed that the particles composing light—photons—sometimes behave more as particles, other times more as waves. They, however, are always both particles and waves at the same time. This is termed the ***dual nature of light***. Several phenomena can be explained considering light as waves, while others are better explained considering light as particles.[1]

If you've ever stood in shallow sea water, you probably noticed that you could not stop the waves with your body. The waves just go around you. This phenomenon is called diffraction, and it lets waves go around obstacles

---

[1]Light behaves as waves in all experiments involving diffraction, interference, reflection, and refraction. It behaves as particles in Einstein's photoelectric effect, Planck's thermal radiation, and Compton's scattering.

Physics in the Arts, Revised Edition

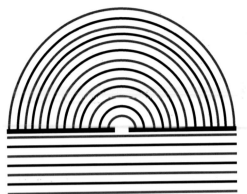

FIGURE 1.1 ▼
Schematic representation of the
waves on water surface, seen from
above. Red lines indicate crests,
black lines troughs. A plane wave
is propagating up from the bottom
of the schematic. When it reaches
a wall with an aperture, it propagates as a circular wave, centered
at the aperture.

almost undisturbed. Waves also go through a small opening and seem to
diverge from there, as shown in Figure 1.1. Also, when the wakes from
two boats combine, standing waves may form. On the water surface, where
both wake waves have a crest, a stationary double-height crest results, but
if both have troughs, there is a stationary doubly deep trough. In places
where one wake wave has a crest and the other has a trough, however, the
water surface is flat, and no wave is present. This phenomenon is called
interference. Constructive interference generates doubly intense crests or
troughs; *destructive interference* cancels a trough with a crest. If you now
consider two slits, producing two sets of circular waves, you will see a similar interference effect. In Figure 1.2, where two black (troughs) or two red
(crests) wave lines intersect, there is constructive interference (solid dots),
and where one red and one black line intersect, the interference is destructive
(open dots).

In a famous experiment done by Thomas Young in 1803, a very similar
interference phenomenon takes place, using light instead of waves on the surface of water and two light beams. Young's experiments proved for the first
time that light is composed of waves. In his experiment, sunlight was reflected
by a mirror and then filtered through a red filter, transmitted through a pinhole,
and projected into a darkened room. At this point the red light beam was split
into two light beams by a card, one-thirtieth of an inch in thickness, held
into the beam edgewise as shown in Figure 1.3. The two beams interfered
constructively and destructively, as in Figure 1.2, and generated the pattern
shown in Figure 1.3.

Often this experiment is thought to have been done by Young using two
slits. That version of the same experiment, however, was not done until much
later. The experiment of Figure 1.3 can easily be reproduced by the reader
using a laser pointer as the illuminating source, or sunlight if the sun is shining,

FIGURE 1.2 ▶ Schematic representation of the same waves going through two apertures and interfering. Where two red lines cross, a crest of double height is formed, while at black line intersections, troughs of double depth are formed by constructive interference (solid dots). Where a black line crosses a red line, destructive interference makes the waves cancel each other, and the water surface is flat (open dots).

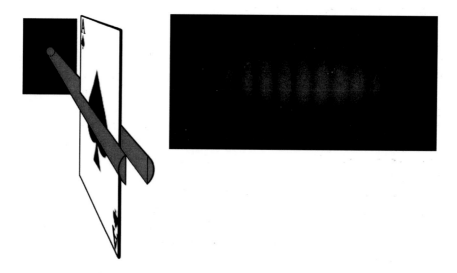

FIGURE 1.3 ▶ Young's experiment demonstrating the wave nature of light. Left: A beam of sun light filtered through a red filter (not shown) is made to pass through a pinhole. A playing card is held edgewise, so the red beam is split into two beams, which interfere after the card to create the pattern on the right.

and a darkened room. It is important to use a colored filter in the latter case; light of only one color (wavelength) must be used.

There is no conceivable way that would make particles accumulate more in certain spots and less in others to generate the interference pattern of

Figure 1.3. Young's experiment, therefore, proved for the first time that light indeed behaves as waves.

In this and all other experiments in which light is better interpreted as waves, there is a similarity between light and the crests and troughs of waves on the surface of water or waves traveling along strings. In light, what oscillates are the *electric and magnetic fields*. They both oscil late in the plane perpendicular to the travel direction. We all have some familiarity with these fields. The magnetic field is what makes a magnet attract a piece of iron, and the electric field is responsible for the attraction of small pieces of paper to a comb that has been rubbed on wool (or dry hair).

As these fields oscillate in light waves, there is an alternation of places with a strong upward electric field, a strong downward field, and an electric field of zero. The pattern is similar to the one we see for waves on water: Light moves with a velocity of propagation v, and, like all other waves, it has an amplitude, a frequency, and a wavelength. Figure 1.4 graphically shows these quantities in a wave.

The wavelength λ varies for different colors. For example, blue light has a shorter wavelength than red light, as shown in Figure 1.5.

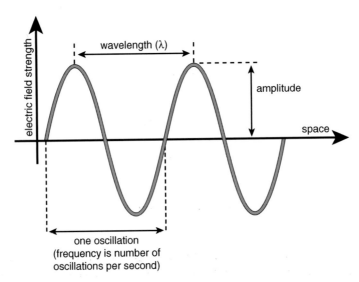

FIGURE 1.4 ▶ Light wave oscillating in space as it travels. The half-height is called amplitude, the distance between subsequent crests is the wavelength indicated by the Greek letter λ (lambda), and the frequency is the number of oscillations in 1 second.

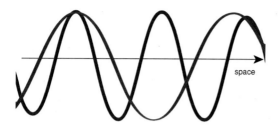

FIGURE 1.5 ▼
Two light waves oscillating as they travel. The amplitude is the same for both waves, while the wavelength for blue light is shorter than for red light.

space

1.1   Speed of Light

The velocity of propagation of light depends on the medium through which it travels. *In vacuum*, and to a good approximation in air, the speed of light is 300,000,000 m/sec (or 3×10^8 m/s). This is a very considerable speed. By comparison, a fast car travels at 30 m/s, sound travels at 300 m/s, and a rocket leaving earth travels at only 10,000 m/s. The speed of light *in vacuum* is usually represented by the letter c (from the Latin word *celeritas*), so we write $c = 3 \times 10^8 m/s$.

In other transparent media, such as water, glass, and plexiglass, the speed (represented as v) of light is smaller than its speed in vacuum c. The actual value of v depends on the medium. The velocity v of light in a medium depends on *the index of refraction n* of the medium and can be found using the following equation:

$$v = \frac{c}{n}$$

The denser the medium, the larger its index of refraction and the slower light travels through it. For example: the index of refraction of plexiglass is $n_{plexiglass} = 1.5$, so the velocity of light in plexiglass is

$$v = \frac{c}{n} = \frac{3 \times 10^8}{1.5} = 2 \times 10^8 \, m/s$$

The index of refraction of vacuum is $n_{vacuum} = 1$, the smallest index of refraction that exists. Therefore, light travels fastest in vacuum. Here is an interesting effect: Imagine a transparent object with $n > 1$ surrounded by vacuum. When light traveling in vacuum traverses an object made of a transparent medium, it slows down, and then it instantaneously accelerates again when exiting the medium, without changing its energy.

1.2 Electromagnetic Spectrum

Electromagnetic waves can have a large range of frequencies or wavelengths. What we call light is the narrow range of wavelengths that can be detected by our eye: **400 nm to 700 nm** (*1 nm = one nanometer = $1 \times 10^{-9}m$*). This range of wavelengths is called *visible range*.

Electromagnetic waves with large wavelengths ($\lambda = 300$ m) are called radio waves and are received by radios. Waves with very short wavelengths are called ultraviolet light, or if shorter still are called x-rays and gamma-rays. Figure 1.6 shows energy, frequency, wavelength, and the common names for the entire electromagnetic spectrum.

Table 1.1 shows the wavelengths corresponding to the seven colors commonly associated with the visible range. These are the colors originally identified by Newton, as dispersed by a prism. There are millions of colors in the spectrum, not only seven, but Newton, by analogy with the seven notes of music, arbitrarily subdivided the spectrum into these seven colors. In spoken language these are the color names most commonly used; therefore, we think it's useful to identify them here by their wavelength.

This could therefore be considered a "spoken language spectrum." In reality the spectrum is a continuum, and a range of wavelengths is associated with each of these colors and a myriad of visibly distinct colors. Some of these—for example, cyan (480–500 nm)—do not normally appear in the "spoken language spectrum," but they are very evident when one looks at the continuum spectrum dispersed by a prism.

EM spectrum

Gamma rays	X-rays	UV VIS Infrared Microwaves Radio waves

Wavelength (m) $10^{-11}\ 10^{-10}\ 10^{-9}\ 10^{-8}\ 10^{-7}\ 10^{-6}\ 10^{-5}\ 10^{-4}\ 10^{-3}\ 10^{-2}\ 10^{-1}\ 1\ \ \ 10\ 10^2\ 10^3\ 10^4\ 10^5\ 10^6$

Frequency (Hz) $10^{20}\ 10^{19}\ 10^{18}\ 10^{17}\ 10^{16}\ 10^{15}\ 10^{14}\ 10^{13}\ 10^{12}\ 10^{11}\ 10^{10}\ 10^9\ 10^8\ 10^7\ 10^6\ 10^5\ 10^4\ 10^3$

Energy (eV) $10^6\ 10^5\ 10^4\ 10^3\ 10^2\ 10^1\ 10^0\ \ \ 10^{-1}\ 10^{-2}\ 10^{-3}\ \ \ 10^{-4}\ 10^{-5}\ 10^{-6}\ 10^{-7}\ 10^{-8}\ 10^{-9}\ 10^{-10}\ 10^{-11}$

wavelength (λ) →

← **energy & frequency (ν)**

FIGURE 1.6 ▶ The electromagnetic spectrum, expressed in wavelengths, frequencies and energies. Notice how narrow the visible range (VIS) is, compared to the full spectrum. The highest energy gamma rays we observe in the universe are 10^{18} eV.

TABLE 1.1 ►

Color Name	Wavelength Range
violet	400–430 nm
blue	430–480 nm
cyan	480–500 nm
green	500–570 nm
yellow	570–585 nm
orange	585–620 nm
red	620–700 nm

1.3 Polarization

Let us now examine one more phenomenon that can only be explained by taking into account the wave nature of light. When light propagates in one direction, the electric and magnetic fields oscillate in a second and third direction, respectively, perpendicular to each other. The plane in which the electric field oscillates, which also contains the direction of propagation, is called the *plane of polarization* (see Figures 1.7 and 1.8). The latter in particular will clarify how the direction of propagation is contained in the plane of polarization. For a simple example, imagine a rope fixed at one end and being moved at the other end. If the free end of the string is moved up and down periodically, the crests and hollows that form on the string are vertically above and below the position of the undisturbed string. In this case we can say that the waves on the string are *vertically polarized*. If the end of the string is moved horizontally, left to right, the waves are *horizontally polarized*.

To produce an electromagnetic wave, an electric charge is moved up and down on a conductor (antenna). This generates a vertically polarized electromagnetic wave, irradiating in all horizontal directions away from the antenna transmitting the signal. This is easily done by making an electric current flow back and forth along a vertical wire.

Light from the sun or a light-bulb, however, is composed of many waves with different, random planes of polarization. This is called *unpolarized light*.

Sunlight is unpolarized. It contains electric fields oscillating in all directions. When unpolarized light is *reflected by a horizontal surface* (such as a mirror, water surface, car top, road, etc.), the *reflected light is horizontally polarized,*[2] as shown in Figure 1.7.

---

[2]At a specific incidence and reflection angle, called the Brewster's angle, 100% of the reflected light is polarized. In other words, if light strikes the surface of a horizontal medium

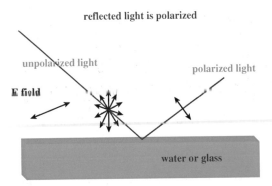

reflected light is polarized

unpolarized light

polarized light

E field

water or glass

FIGURE 1.7 ▼
Unpolarized light has the electric field oscillating in all directions perpendicular to the direction of propagation. When unpolarized light is reflected by a horizontal surface of water or glass it becomes polarized, horizontally polarized.

Sunglasses with polarizing lenses

light waves with E
vibrating horizontally

light waves with
E vibrating vertically

FIGURE 1.8 ► Polaroid filters in sunglasses are always mounted so that horizontally polarized light, that is, light with the electric field E oscillating horizontally, shown in red, reflected by horizontal surfaces (water, car tops, floors, roads etc.), is absorbed, and light polarized in all other directions, or unpolarized is transmitted. Polaroid sunglasses do not stop the glare from windows, as this is vertically polarized, shown in blue.

Polaroid filters absorb light polarized in one specific direction and transmit the rest. These filters are prepared by depositing iodine needle crystals onto a polymer surface, which is then stretched so that the iodine needles orient themselves along the direction of stretching, causing them to absorb light polarized in that direction only.

Sunglasses with polarized lenses have the iodine needles oriented horizontally, and they absorb horizontally polarized light (Figure 1.8). Vertically

at a Brewster's angle, the reflected light is completely horizontally polarized. For water the Brewster's angle is 53°, for glass it is 56°, and for diamond 67.5°. At incidence angles that are larger or smaller, part of the light is also vertically polarized, so the degree of horizontal polarization is less than 100%.

FIGURE 1.9 ▶ Photographs without (left) and with (right) polarizer filter of the same clouds. Notice how the contrast cloud-sky is enhanced by the polarizer. Photographs are courtesy of Tiziana Parasassi.

polarized light is still transmitted. If you combine two Polaroid filters at 90° from each other, no light is transmitted.

A *polarizer filter* in front of a camera lens darkens the clouds in the sky but darkens the sky even more. This happens because light scattered by smaller particles is more polarized than light scattered by larger particles. The air molecules in the sky are smaller than the ice crystals and water droplets in the clouds. Therefore, there is more polarized light from the sky than from the clouds. Correspondingly, the polarizer filter removes more light from the sky, and the contrast between the clouds and the sky is enhanced, as shown in Figure 1.9.

► Chapter 2

Reflection and Refraction

2.1 Specular Reflection of Light

Imagine a mirror surface that is perfectly flat, polished, and reflecting. Now imagine a line perpendicular to the mirror surface, called the *normal*. The normal forms angles of 90° from the mirror surface. When light illuminates any mirror like the one you imagined, it is reflected.

In Figure 2.1, the ray of light coming from the left is incident (or "incoming") on the mirror surface. The angle marked *i* is called the *angle of incidence*, and it is measured from the normal (in the particular diagram of Figure 2.1, $i = 45°$). The incident ray is reflected by the mirror into another ray. This reflected ray of light forms an *angle of reflection*, marked *r*, with respect to the normal. ***The angle of incidence is equal to the angle of reflection***, or

$$i = r.$$

This is called the **law of specular reflection**, from the Latin word *speculum*, meaning "mirror." The law of specular reflection is valid for any value of the angle of incidence *i*, not only for the 45° angle shown in Figure 2.1. This law is all you really need to know about specular reflection.

All polished and smooth surfaces reflect light as a mirror. Usually mirrors are made by combining a glass pane, which can be made very flat and smooth, with a reflective metal layer coated on its back.

All materials reflect light, at least to some extent. This is true even for the most transparent ones, such as uncoated glass. You can see this in everyday

Physics in the Arts, Revised Edition

Specular Reflection: i = r

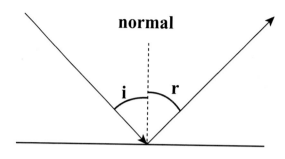

$$i = angle\ of\ incidence$$
$$r = angle\ of\ reflection$$

FIGURE 2.1 ▶   A ray of light is reflected by a mirror so that the angles of incidence and reflection are equal ($i = r$).

life. When you are standing outside in daylight and look at a building, all the windows reflect the outside world, including the sky, other buildings, and trees. Why does this happen?

This happens because the windowpanes are very flat and slightly reflecting. Since there is more light outside than inside the building, the windows reflect the outside world. The same phenomenon happens at night in the opposite direction. When there is artificial illumination inside (the lights are on) and it is dark outside, people from outside can see the inside through the windowpanes very well, but from the inside, people can't see out. If you are standing inside, you can only see the reflection of yourself and the rest of the room. If you want to see outside, you must turn the inside lights off. The general rule is that you can see from darker to brighter through a window but not vice versa. Remember that this is because of reflection. You only see reflection if you stand on the bright side.

Another observation is that the reflectivity of polished surfaces is greater at grazing incidence—that is, when the angle of incidence (as well as the angle of reflection) is close to 90°.

Most objects around us are not perfectly polished, so they are not good specular reflectors. When these objects are illuminated, they do not reflect light according to the law of reflection. Instead, nonpolished objects *scatter* (or *diffusely reflect*) light in all directions around them. This means that you do not have to be in a specific position or at a specific angle to see an illuminated object. You can be at any angle from it and still see it. This is quite convenient!

Every nonpolished, illuminated object can be considered a source of light rays. The sun, incandescent lightbulbs, neon lights, fluorescent lights, or computer monitors all emit light, but not very many other objects do. Illuminated nonpolished objects, however, can be represented as sources of light, since they *scatter* light in all directions. From the geometric point of view, objects can be schematically represented as centers of diverging rays— that is, sources of light rays. Imagine for example a red laser pointer beam illuminating a spot on a sheet of paper. The red dot on paper can be seen from everywhere else in the room, not only at a specific angle, as it would on a mirror. Therefore, light rays diverge from the red dot on paper as if that dot was emitting light. Clearly the paper does not emit light but only scatters it, as most other objects surrounding us do. The dot on the paper, however, can be considered as a source of light rays, propagating in all directions.

Imagine now that an object is in front of a mirror. The diverging rays of light that originate from the object, marked O in Figure 2.2, are reflected by the mirror. The reflected rays of light *behave as if* they originated from the point I. This point is called the **virtual image** of the object O.

Several observations can be made from this figure:

- Reflected rays seem to come from point I.

- Point I is located behind the plane of the mirror, at a distance h, equal to the distance between O and the mirror.

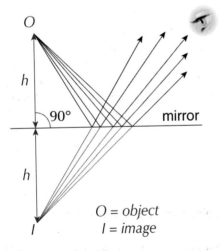

The image is *virtual*, not real

O = object
I = image

FIGURE 2.2 ▶ Reflection from a mirror.

- Points O and I are on a line perpendicular to the plane of the mirror.

- The mirror does not need to extend to the red line on the left. A smaller mirror is sufficient to produce an image of object O.

To find the image position of any object, draw the normal from the object to the mirror and then measure the distance between the object and the mirror. The image I is located at the same distance from the mirror as the object, but it is on the opposite side. This is why it is called a *virtual image*. In other words, **there are no real rays of light that emerge from point I**. As you can see in Figure 2.2, **only the artificial extensions of the reflected rays (marked as dashed lines) originate from I**.

Any ray from O that strikes the mirror is reflected, and the artificial extensions of these reflected rays always seem to come from point I. If you were to remove the mirror and place point object O at point I, it would look identical to the reflected point O. The eye cannot tell the difference between rays coming from a virtual image I and a real object at the same position.

When you look at yourself in the mirror, what you see is a virtual image of yourself. This virtual image is behind the mirror. If a friend were to look behind the mirror, however, she would not find an image of you. The image is virtual!

You can only see a virtual image if you are on the same side of the mirror as the object.

Another important observation is that vertically mounted mirrors appear to invert left and right, but not top and bottom. Both these observations are actually incorrect. The vertical mirror does not invert left to right but only front and back. In the mirror on the wall you see the top of your face on the top and your right ear on the right. Neither up-down nor left-right are reversed. The only inverted direction is front and back. Other people facing you, however, always see your right ear on their left, so they see you differently from the way you see yourself in the mirror. This is why everybody thinks that left and right are inverted in a vertical mirror. It's just the habit we all have of seeing each other frontally.

Why is gold yellow? Why are silver and aluminum gray? Why is copper red? In Figure 2.3 we report the **reflectivity curves** for these metals. As mentioned in Chapter 1, the wavelengths between 400 nm and 700 nm are the **visible range** of light. Below 400 nm is ultraviolet light, and above 700 nm is infrared light. We do not see these types of light. From the reflectivity curves, you can see that gold does not reflect blue. Rather, it absorbs blue and this causes it to appear yellow. Copper absorbs both blue and green, so it looks orange-red. Aluminum and silver have fairly flat reflectivity curves along the entire visible range, so they appear gray. In other words, aluminum and silver reflect all the colors of light illuminating them and absorb none. This is why silver and aluminum are the best materials for mirrors. In the old days, silver was the only material used to make mirrors, so few mirrors were produced, and they were very expensive. Today, technology is available that allows us to deposit thin, smooth layers of aluminum on glass. Mirrors made through this

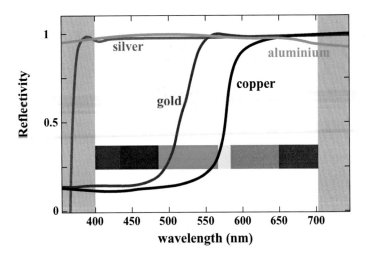

process are much more affordable, so now mirrors are much more widely available. Can you imagine life without a mirror? If not, you owe a thanks to physics!

2.2   Refraction of Light

Light travels along straight lines in a uniform medium. If a ray of light passes from one medium to another (from air into water or glass, for instance), in a direction that is normal, or perpendicular, to the interface, the ray will travel in a straight line. However, if a ray passes from one medium to another *in a direction different from the normal*, the direction of the light ray changes in the second medium—*the ray is kinked*. This abrupt deflection, or kinking, of light rays at the interface between two media is called ***refraction***.

The behavior of light in a medium depends on the ***index of refraction*** of the medium, n. The index of refraction of a medium determines the speed of light in that medium and the angle of refraction. The index of refraction n is an intrinsic property of each material and depends on its density. In general, the denser the medium, the higher the index of refraction n. Vacuum, and to a good approximation air, has an index of refraction $n = 1$. For water, $n = 1.33$, and for glass, $n = 1.52$. The index of refraction n is a dimension-less number—in other words, it has no units. In most media, n is a number between 1.0 and 2.5. Table 2.1 presents the indices of refraction of common materials.

TABLE 2.1 ▶

Material	Index of Refraction n
vacuum	1.00000
air at 50°C	1.00024
air at 0°C	1.00029
liquid water (20°C)	1.33
ice	1.31
ethanol	1.36
plexiglass	1.50
glass (crown)	1.52
glass (flint)	1.89
zircon	1.97
cubic zirconia	2.16
diamond	2.42
gallium phosphide	3.50
gallium arsenide	3.93
silicon	4.01

In the real world, things do not work out so simply. There is an additional complication: The index of refraction is not constant for each material but actually depends slightly on the wavelength of light. This phenomenon is called *dispersion*. For example, the index of refraction of borosilicate crown glass (known also as BK7) for short 400 nm wavelength (violet) is $n_{400} = 1.53$ and for long 700 nm wavelength (red) is $n_{700} = 1.51$. This may seem a small difference, but as we will see later on, it makes it possible to separate the spectral colors using a crown glass prism.

Let us consider a light ray crossing the interface between a medium with an index of refraction $n = 1.0$ and a second medium with an index of refraction $n > 1$. This is depicted in Figure 2.4. Notice in Figure 2.4 that a ray of light

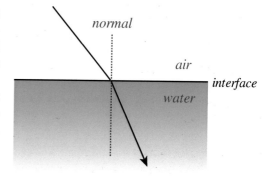

FIGURE 2.4 ▶ Light traveling from air to water. When it traverses the water surface, the light ray is kinked toward the normal.

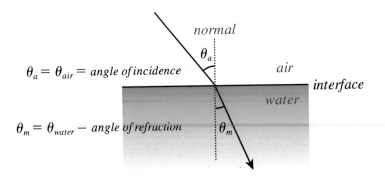

FIGURE 2.5 ▶ Definitions of the angles of incidence and refraction.

is traveling in air and then reaches the interface between air and water. As the ray crosses the interface and enters the water, it is kinked toward the normal. Whenever light travels from a less dense to a denser medium, it kinks *toward the normal*, as shown in both Figures 2.4 and 2.5. When light travels in the reverse direction from a denser to a less dense medium it kinks *away from the normal*.

Let us compare the angle that a ray forms with the normal at the point of incidence (θ_a, the angle in air) and the corresponding angle in water (θ_m), shown in Figure 2.5. The behavior of the ray as it is refracted at the interface can be described as follows:

1. The angle in air (θ_a) and the angle in the medium are on opposite sides of the normal (Figure 2.5).
2. The angle in air is always larger than the angle in the other medium because air is less dense than any other medium (e.g., glass, water, plexiglass, acrylic, etc.).
3. The angles θ_a and θ_m do not change if the light ray goes from air to the medium or from the medium to air. The ray of light is reversible. This is always true. *Light rays are always reversible*. We will make great use of this characteristic of all light rays, here and later.
4. The angles θ_a and θ_m are related to each other in a very accurate and quantitative law—called Snell's law.

$$\text{SNELL'S LAW} \qquad \frac{\sin \theta_a}{\sin \theta_m} = \frac{n_m}{n_a}$$

Refraction: Simplified Snell's Law

$$\frac{A}{M} = n_m$$

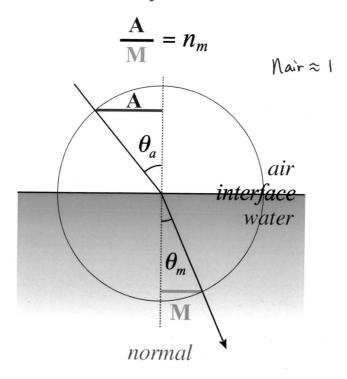

FIGURE 2.6 ▶   A visual representation of Snell's law, in which A is the sine of θ_a and M is the sine of θ_m. These correspond to the blue and green horizontal lines, respectively.

2.3 | Total Internal Reflection

Any angle in air θ_a corresponds to a particular angle θ_m in the medium. For example, for rays traveling from air into water (with an index of refraction $n = 1.33$), Snell's law gives the results shown in Table 2.2.

Table 2.2 goes up to 90° only. This is because the angle in air, θ_a, can never be greater than 90°. If it were greater, the ray would no longer be coming from air but from under the surface of water (below the interface).

Table 2.2 can be also used for a ray of light traveling in the opposite direction, from water into air. A diver with a waterproof flashlight could experiment with rays of light and verify for us that the angles reported in the table are still accurate when going from water to air.

TABLE 2.2 ▶

Angle in Air θ_a	Angle in Water $\theta_m = \sin^{-1}(\theta_a/n)$
0°	0°
10°	7.5°
20°	14.9°
30°	22.1°
40°	28.9°
50°	35.2°
60°	40.6°
70°	44.9°
80°	47.8°
90°	48.7°

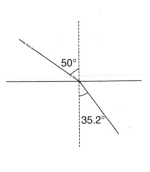

Remember, *rays of light can always be inverted*. All optics formulas and all drawings, including Snell's law and the table of values for θ_a and θ_m, do apply to inverted rays. When using inverted rays, the only difference for Snell's law is that you start from the angle in water and calculate the angle in air. Looking at Table 2.2, you would start from the right column and read the corresponding angles on the left. There is one thing you must watch out for when using inverted rays: Since $n_{water} > n_{air}$, the angles in water θ_m do not go all the way to 90°. Notice: the table does not tell us what happens with a ray with $\theta_m = 60°$. What happens is simply the following:

$$\theta_m = 49° \text{ corresponds to } \theta_a = 90°$$
$$\theta_m = 60° \text{ corresponds to } \theta_a > 90°$$

This is *impossible* because this ray would not really be in air. The ray is not refracted into air but **reflected** at the interface of water-air back into water!

Above a certain angle, called the **critical angle** θ_c or, more precisely, the critical **angle of total internal reflection**, all light is reflected from the interface. No light is refracted. This occurs only when the ray of light passes from a denser to a less dense medium—that is, from higher to lower index of refraction. For example, this can happen when a ray is passing from water to air, as just described, but not from air to water.

The critical angle for the water/air interface is 49°. For glass/air it is 41° and for diamond it is 24.4°.

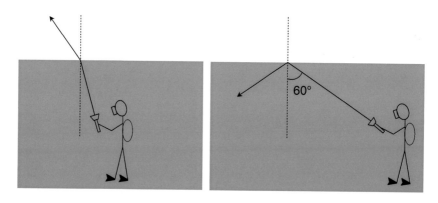

FIGURE 2.7 ▶ A man under water can project light into air from his flashlight, only as long as the incidence angle is less than 49°, the critical angle of total internal reflection for the water/air interface.

This means that if the angle of incidence is greater than the critical angle, $\theta_m > \theta_c$, *no light is refracted*. In Figure 2.7, since the angle of incidence is 60°, and it is greater than the critical angle $\theta_c = 49°$, the light is totally reflected.

It is easy to calculate the critical angle. At the critical angle, the angle in air $\theta_a = 90°$. Since we know that the index of refraction n for air $= 1$ and that $\sin 90° = 1$, Snell's Law appears as

$$n_{water} = \frac{1}{\sin \theta_{c(water)}}$$

Hence,

$$\sin \theta_{c(water)} = \frac{1}{n_{water}}$$

from which we can calculate.

$$\theta_{c(water)} = \sin^{-1}\left(\frac{1}{n_{water}}\right) = \sin^{-1}(0.75) = 49°.$$

With the aid of the preceding information, we can anticipate the behavior of a light ray as it crosses a transparent medium. This allows us to make drawings of what will happen without actually doing experiments!

Figures 2.8, 2.9, 2.10, and 2.11 show examples of refraction in some common situations. In all of these examples, for simplicity's sake, we only show the rays coming from a few points of the object. Be aware that there are similar rays coming from a few points of the object. You can imagine what these rays would look like.

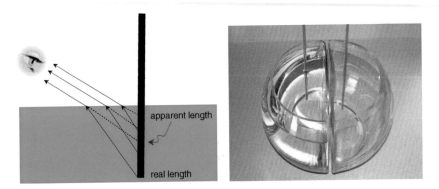

apparent length

real length

FIGURE 2.8 ▶ Because of refraction, a vertical stick under water appears to be shorter. If the stick were not vertical, it would also appear to be bent at the air-water interface. The bottom of the stick always appears closer to the surface than it really is. The photograph shows two identical bowls (approximately semispherical) and two turquoise glass rods, which are vertical and identically positioned in the two bowls. The left bowl is filled with water, and the right bowl is empty. Notice the striking difference in apparent length: The underwater rod appears shortened by refraction.

FIGURE 2.9 ▼
A fish under water seen from the top, appears to be closer to the surface than it really is. The apparent size of the fish, however, does not change.

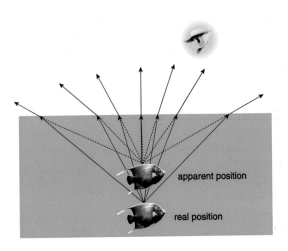

apparent position

real position

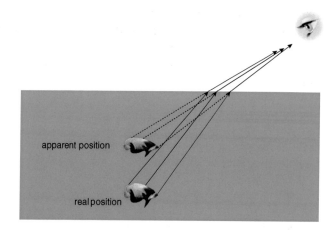

FIGURE 2.10 ▶   A fish seen at an angle (not from the top) appears to be closer to the surface and more vertically compressed (narrower) than it really is. Experienced spear fisherfolk keep refraction into account when they aim and launch their spear.

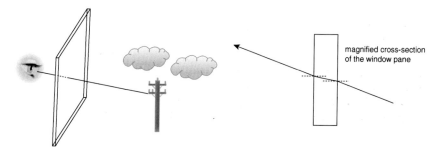

FIGURE 2.11 ▶   Here a ray crosses a glass pane with parallel sides. The ray coming from the right is kinked toward the normal (dashed line) when entering the glass and away from the normal when exiting on the left. The incoming and outgoing rays are parallel to each other. Since the windowpane is thin, there is almost no shift in the image position. A thicker block of glass or plexiglass shifts light rays much more than a windowpane does.

2.4   Reflection and Refraction in Diamonds

A brilliant cut diamond sparkles because of skillful use of refraction and total internal reflection. Let us look at the detailed path of some rays of light. Consider two rays of light (1 and 2 in Figure 2.12) passing the air-diamond interface. Ray 1 is kinked toward the normal (red), because a diamond is denser than air. Ray 2 enters perpendicular to the surface of the diamond (along the normal) and therefore does not kink. Instead, it goes straight through the surface. Both rays reach the inner surface at the arrow heads and from there undergo total internal reflection. The angles that fall within the critical angle are shown by the blue cones in Figure 2.12. Both rays 1 and 2 are *outside* these blue cones and are thus both at angles larger than the critical angle. Both rays are, therefore, totally internally reflected. The critical angle is calculated, as before, using Snell's Law. The index of refraction for diamond is $n_{diamond} = 2.42$.

Snell's law for diamond is

$$\frac{\sin \theta_{air}}{\sin \theta_{diamond}} = \frac{n_{diamond}}{n_{air}}$$

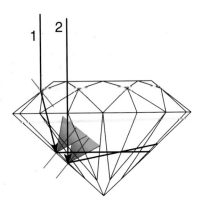

FIGURE 2.12 ▶ Rays 1 and 2 are totally internally reflected because they reach an inner diamond surface at angles greater than the critical angle, which for diamond is very small: only 24.4°.

Total internal reflection in diamond takes place when $\theta_{air} = 90°$, which corresponds to

$$\sin \theta_{diamond} = \frac{n_{air}}{n_{diamond}} = \frac{1}{2.42} = 0.41 \text{ and } \sin^{-1} 0.41 = 24.4°$$

Whenever $\theta_{diamond} \geq 24.4°$, light is completely reflected within the diamond and no light is refracted outside.

After rays 1 and 2 are totally internally reflected, the two rays reach the other side of the diamond bottom surface. Here, they are again both traveling at angles greater than the critical angle (outside the blue cones), as shown in Figure 2.13. Therefore, they are completely reflected toward the top of the diamond.

Ray 1 is reflected in a vertical direction and reaches the top surface perpendicularly. Therefore, it goes straight through the surface. Ray 2 reaches the surface at an angle. This

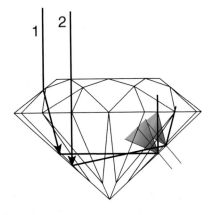

FIGURE 2.13 ▶ Rays 1 and 2 again.

time, the incidence angle is *within* the blue cone, so the ray is refracted outside the diamond. Since the index of refraction n is slightly different for each light wavelength, each color of the spectrum is refracted at a slightly different angle as shown in Figure 2.14. Violet rays are kinked more than red rays. In physics, this phenomenon is called *dispersion of light*. In the jewel industry, it is called the *fire* of a gemstone. Light dispersion will be discussed in more detail in Chapter 6.

If the diamond is cut too shallow or too deep (as shown in Figure 2.15), light leaks out through the bottom facets, and the diamond appears dark and dull. The cones defining the critical angle are again shown in blue. Notice that where the light leaks out, the rays are traveling at an angle that falls within the blue cones. If light is able to "escape" in this way, a diamond does not appear as sparkly.

For the diamond to sparkle, it must be cut at very accurate proportions— for instance, the "brilliant cut" and

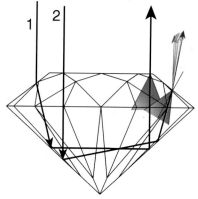

FIGURE 2.14 ▶ Light of different colors is dispersed at different angles, generating the "fire" in a diamond.

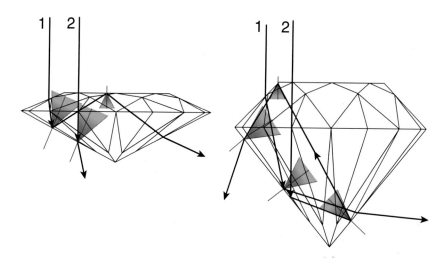

FIGURE 2.15 ▶ Two diamonds, one cut too shallow and one too deep, are shown here. The cones defining the critical angles are again shown in blue. Notice where the light leaks out, the rays are traveling at an angle that falls within the blue cones.

its proportions, optimized to give the greatest sparkle to a diamond, are shown in Figures 2.16 and 2.17.

After cutting the gem, the surface of the diamond facets must also be polished. If they are not polished, the angles of total internal reflection are not all oriented equally with respect to the surface, as shown in Figure 2.18. This causes most of the light to be lost to refraction.

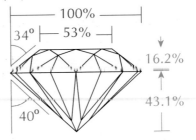

proportions of a brilliant cut diamond

FIGURE 2.16 ▶ The precise proportions of the "brilliant cut" for diamonds.

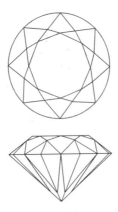

FIGURE 2.17 ▶ Top view of the circular crown (top) and side view of a brilliant cut diamond (bottom).

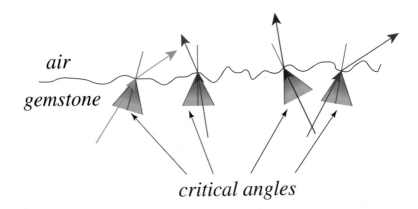

FIGURE 2.18 ▶ Diamond facets must be polished or most of the light will be lost due to refraction within the critical angle cones.

2.5   The Rainbow

Now that we have familiarized ourselves with *reflection, refraction,* and *dispersion,* we can understand how rainbows form. In general, to be able to see a rainbow in front of you, the sun must be behind you. A myriad of water droplets, such as rain or moisture, must be present in the air for the rainbow to appear. This is shown in Figure 2.19. Three of the droplets are magnified there. The black lines represent light rays coming from the sun. Droplet 1 shows how the white light from the sun is *refracted* inside the droplet (toward the normal) and then *dispersed* so that violet rays are deflected more than red rays. All of the light rays are then partly *reflected* on the inner surface of the droplet, following the law of reflection. Notice that this is *not* total internal reflection! In fact, light rays are partly reflected and partly refracted outside the droplet (away from the normal). The reflected rays then reach the droplet surface again, and, again, they will be partly reflected (see droplet 1 in Figure 2.19) and partly refracted outside the droplet.

Droplets 2 and 3 show the red and the violet light rays only. Hopefully this will clarify that there are different angles and directions of refraction and internal reflection for each color wavelength. In reality, all colors come out of all droplets at all times, so it is incorrect to assume that 2 is a "red droplet" and

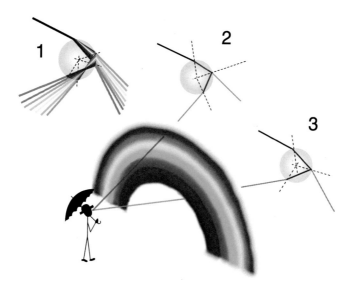

FIGURE 2.19 ▶   How the rainbow forms. Each water droplet, schematically represented by a sphere, is penetrated by white light, which is dispersed, then partly refracted outside the droplet, partly internally reflected.

that 3 is a "violet droplet." They only appear to have those colors to the little man with the umbrella because these droplets are at a well-defined position in the sky. This allows the refracted-reflected-refracted rays to reach his eyes at the correct angle to appear red and violet, respectively. Notice how the violet rays come out of the droplets at a shallower angle, while the red rays come out at a steeper angle. Likewise, red always lies along the outside arches of the rainbow, while violet is always along the inside. So droplet 2 is above droplet 3 in Figure 2.19.

A straight line from the sun through the eye of the man with the umbrella and continuing on marks the center of the rainbow. In this sense the man with the umbrella carries the rainbow around with him. Sometimes on top of a mountain or from an airplane one can see almost an entire rainbow circle.

On rare occasions **double rainbows** can be observed in nature. The second rainbow always lies outside the first one and always has an inverted color order (violet outside, red inside). The second rainbow is formed by the second internal reflection and the fourth refraction outside of the droplet that follows. The droplet in Figure 2.20 shows a schematic representation of the first four refractions.

Notice that at each refraction some of the light is lost in a direction that cannot be observed from ground. Therefore, the second rainbow is always dimmer than the first. At this point it may be a bit difficult to imagine how these two rainbows can form at once. Figure 2.21 shows a droplet that could "potentially" generate both rainbows. In reality, no droplet alone can generate either a single or a double rainbow. You have to imagine a plethora of droplets.

Those droplets that are at the correct position for an observer to see them as red will appear red. The next ones will appear orange, the next yellow, the next green, and so forth. Observe how the doubly reflected rays that form

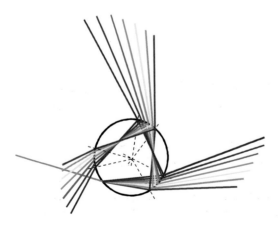

FIGURE 2.20 ▼

The first four refractions of dispersed rainbow colors are shown here. The first occurs when white light enters the droplet, the following ones when the dispersed colors are refracted outside the droplet. Each refraction is less intense than the previous.

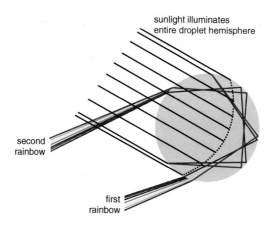

FIGURE 2.21 ▶ The origin of the first and the second rainbow. The second rainbow is always much less intense than the first, because light from each second-rainbow droplet has already undergone two dispersive refractions (as shown in Figure 2.20), and has lost intensity.

the second rainbow have an inverted color order with respect to the first rainbow. Inside the droplet of Figure 2.21, again for simplicity's sake, only violet and red rays are shown. The final refracted rays (the rainbows), however, are depicted as the seven commonly mentioned colors of the spectrum. Be aware that this is only a simplification. There are not really only seven colors in a rainbow but an infinite number of colors all associated with the infinite array of continuously increasing wavelengths along the entire visible light spectrum. In this range, the human eye can distinguish approximately one million colors. Water droplets disperse all different wavelengths with continuity, along with the seven colors that are commonly associated with a rainbow. No abrupt separations between colors can actually be observed in the rainbow or any other dispersed spectrum.

2.6 Questions

1. Is the image you see of yourself in the mirror real or virtual? Explain.
2. Which way is a beam of light kinked when it enters water from air?
3. When it goes from water to air?
4. For spear fishing you have to look straight down to the water. Why? Where would the spear go, with respect to the fish, if you aimed for the fish, looking at an angle other than 0° from the normal? Similarly, explain why a person standing in water looks out of proportion to a person outside the pool. What does the person in the water look like?

5. If you look at yourself in the mirror, do you see yourself identically to how other people normally see you, leaving psychology aside? Are left and right reversed? Are up and down reversed? Why?

6. How can one observe dispersion?

7. What is *dispersion*?

8. Define *index of refraction*, *Snell's Law*, and *total internal reflection*.

9. You stand 4 meters from the mirror. Your friend stands 2 meters from the mirror. To *you*, where does the image of your friend appear to be?

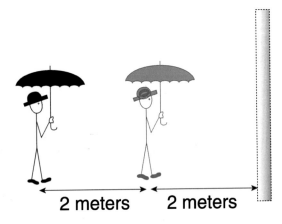

2 meters   2 meters

Side view

10. In the Top View diagram below, you are the person with sunglasses, looking in the mirror at the image of your friend. Carefully draw the path of light rays from the tip of your friend's nose to your left eye and to your right eye. Extend the two rays that go to your eyes to illustrate where they seem to come from.

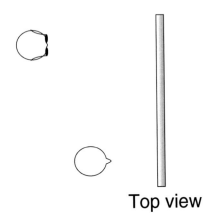

Top view

11. Show by rays reflected from a mirror where the mirror image of the flower appears to be.

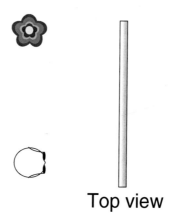

Top view

12. If you are 6 feet tall, how tall a mirror do you need to see yourself head to toe? Where on the wall should the mirror be mounted? Show a diagram to explain your answer.

13. The index of refraction of transparent acrylic is $n = 1.414$. What is the critical angle for this material?

14. The index of refraction of crown glass is $n = 1.52$. What is the critical angle for this material?

▶ Chapter 3

Lenses

3.1 The Prism

When the surfaces of a block of glass or plexiglass are not parallel, as in a prism, the incoming and outgoing rays of light are not parallel. In Figure 3.1, the incoming ray is first refracted and kinked toward the first normal. Then it is transmitted through the prism, and when it reaches the other surface, it is refracted again, only this time away from the second normal. When light travels through a prism, the end result is *always* that the ray is deflected *away from the apex* of the prism.

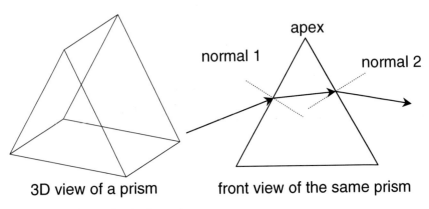

FIGURE 3.1 ▶ Refraction by a prism.

3.2 Converging and Diverging Lenses

Converging lenses are *thicker* in the middle than near the edges, so they *deflect light rays toward an axis*. You can see this if you apply the law of refraction to the surface where the light beam enters the glass and to the surface where the light beam comes out of the glass to find the direction in which light will kink with respect to the normal. You may think of a little piece of the lens as similar to a prism, as shown in Figures 3.2 and 3.3. Figures 3.4 and 3.5 show a few light rays going trough a converging and a diverging lens, respectively.

Figure 3.6 shows some other lens shapes. The more curved the surfaces, the more the rays are deflected as they travel through the lens. The higher the refractive index of the lens material, the more the rays are deflected.

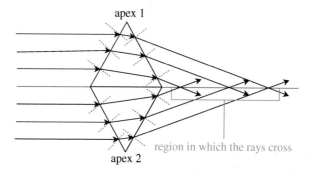

FIGURE 3.2 ▶ Combination of two prisms. The top prism deflects the rays down, while the bottom prism deflects them up. The rays that travel through equivalent sections of the two prisms cross each other at the center, on the pink axis. But is there a transparent object that can make *all* parallel rays that travel through it cross at the exact same point, no matter where they traverse the object?

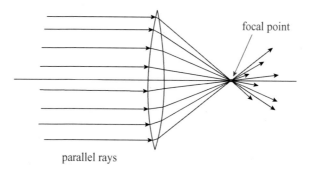

FIGURE 3.3 ▶ The answer to the question in Figure 3.2 is yes! If the two surfaces of the transparent prisms are not flat but curved along two spherical surfaces, the result is a lens. In the diagram here, the lens is convex on both sides, so it is a converging lens. All parallel rays coming from the left *converge* into a single point on the right!

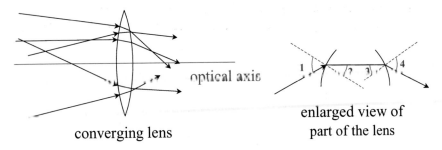

optical axis

converging lens

enlarged view of
part of the lens

FIGURE 3.4 ► Notice in the enlargement that, in accordance with Snell's Law, angle 2 is smaller than angle 1, and angle 4 is larger than angle 3.

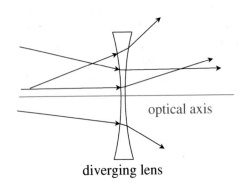

FIGURE 3.5 ▼
Diverging lenses are *thinner in the middle*, so they deflect light rays *away* from the axis.

optical axis

diverging lens

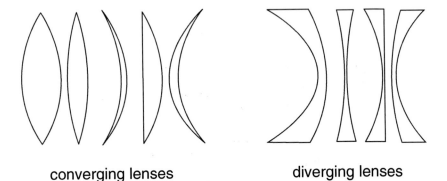

converging lenses

diverging lenses

FIGURE 3.6 ► The concavity or convexity of lens surfaces is not sufficient to determine if a lens is converging or diverging. The real distinction is the thickness at the center of the lens compared with its edges: converging lenses are thicker in the middle, diverging lenses are thinner in the middle.

Eyeglasses are diverging or converging, depending on whether the person is nearsighted or farsighted. If you wear glasses, check to see if they are thicker or thinner at the outer edges.

3.3 Focal Length

Consider parallel light rays (for instance light rays from a light source very far away or at infinity) falling upon a converging lens.

As shown in Figure 3.7, the rays converge at a point at a certain distance f from the lens. This is called *the focal length* of the lens. Conversely, if you place an object (a source of light) at distance f from the lens, the light rays will pass through the lens and come out parallel to each other (light rays can always be inverted). A *strong lens* is one that deflects light a lot, meaning it has a *short focal length*. The focal length of camera lenses is always given in millimeters. The most common and inexpensive 35 mm cameras (indicating 35 mm film width, *not* 35 mm focal length!) have lenses with a focal length of 50 mm ($f = 50$ mm). For eyeglasses, the *dioptric power* is usually used instead of the focal length to describe the lens. The relationship between dioptric power (measured in diopters, D) and f is simple:

$$\text{Dioptric Power (D)} = \frac{1}{f(m)}$$

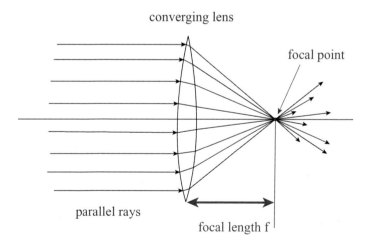

converging lens

focal point

parallel rays

focal length f

FIGURE 3.7 ▶ When parallel rays are refracted by a converging lens they cross each other in the focal point. The distance of this point from the center of the lens is called focal length.

The focal length f in this formula *must* be expressed in *meters*! If you forget, it is like mislabeling a piece of mail. If you put the wrong house number, the letter may still get to the right street but not the right house. Similarly, the diopters will be off by several decimal places. In physics, that's a big difference!

According to the formula, a camera lens with a focal length $f = 50 \text{ mm} = 0.05 \text{ m}$ has $\dfrac{1}{0.05} = 20 \text{ D}$. The advantage of diopters is that if you combine two lenses, one right next to the other, you can simply add the diopters. This implies that for two lenses, the combined focal length is

$$\frac{1}{f} = \frac{1}{f_1} + \frac{1}{f_2},$$

where f_1 and f_2 are the focal lengths of the two lenses.

After the parallel rays go through the diverging lens, they diverge *as if* they came from the virtual image at the focal point. The effect of a diverging lens is shown in Figure 3.8. Just like with converging lenses, the focal point is found at a distance f, or the focal length, away from the lens. To indicate that the (apparent) crossing point of the rays is on the same side of the lens as

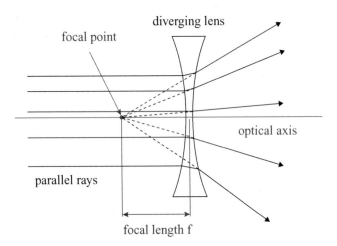

FIGURE 3.8 ▶ A diverging lens with parallel rays incident upon it from a faraway object. The diverging lens makes the rays diverge, as if they were originating from the focal point. The image of the far away object (at infinity) on the left is formed in the focal point on the left-the same side of the lens as the object. ***The image is virtual,*** as there are no actual rays converging at the image point. Only the artificial extension of diverging rays (shown as dashed lines in this diagram) converge in the focal point or, more accurately, *diverge from the focal point.*

the source of light, ***the focal length of a diverging lens is stated as a negative number***. An important advantage of giving the focal length of a diverging lens as a negative number is that it makes the formulas work out automatically. For instance, if you combine a converging lens with 10 cm focal length with a diverging lens of -10 cm focal length, the resulting focal length, f is calculated from the following equation:

$$\frac{1}{f} = \frac{1}{10} + \left(\frac{1}{-10}\right) = \frac{1}{10} - \frac{1}{10} = 0$$
$$f = \infty$$

In the system of two lenses shown in Figure 3.9, the resulting focal length is infinite—the incoming rays are parallel, and the outgoing rays are also parallel. The lens combination acts like a flat windowpane! But wait—does this really work? Yes, because when a converging lens and a diverging lens are combined so that the convex and concave sides are stuck together, the resulting object is indeed a parallel slab of glass, just like a windowpane!

The focal length of a lens depends on the index of refraction of the glass and on the radius of curvature of the two faces of the lens, r_1 and r_2, as shown in Figure 3.10:

$$\frac{1}{f} = (n - 1) \left[\frac{1}{r_1} + \frac{1}{r_2}\right]$$

This is called the *lens-maker's formula*. This formula has important consequences that are worth noticing.

There are always two focal points for each lens, and they lie at a distance f on either side of the lens.

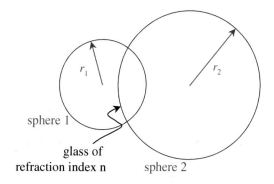

FIGURE 3.9 ▶ System of 2 lenses, which combined have infinite focal length.

FIGURE 3.10 ▶ A lens with two different radii of curvature, as in the lens maker's formula.

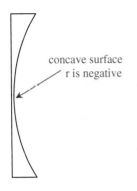

concave surface
r is negative

FIGURE 3.11 ▶ If the lens surface is concave, the radius is negative.

The lens may have two different radii of curvature, but the focal points are always equidistant from the center point of the lens.[1] This implies that a lens can be used from one side or flipped and used from the other side, and either way the image will be identical. If you wear glasses, you can try this right now! Take your glasses off, rotate them by 180° so that you can look through the lenses and the arms extend away from your face. Now look through the lenses. If you rotated the glasses correctly, you should be looking with your right eye through the right lens and with the left eye through the left lens—but inside out. You should be able to see just as well with the lenses inverted. Taking into account how the lenses are curved, this is really counterintuitive! It is one of many surprising phenomena in physics.

The formula makes sense even if $n = 1$ (a lens made of air!), is always zero, or $f \to \infty$, and the rays are not deflected.

If the face of the lens is **concave**, r is counted **negative**, as shown in Figure 3.11.

The smaller r_1 and r_2 get, the smaller the focal length. What does the formula tell you if you take a thin sheet of plastic and curve it? Does it still make sense?

Note that the lens maker's formula **works only for thin lenses**—lenses for which the thickness is much less than the focal length.

3.4   Images—Real and Virtual

Imagine sunlight shining on the face of a lion. Think of one tiny point on the lion's cheek. Part of the sunlight hitting that point is **scattered in all directions.** That's why we see the lion with our eyes no matter from what direction we look at it! This is also called diffuse reflection of light, and it is different from specular reflection from a mirror, which goes in one particular direction only. If a converging lens is placed in such a way that the rays coming from the point of the lion converge to a point on the other side of the lens, an image of the object point on the lion is formed. Figure 3.12 shows this case. The lion is

---

[1] This is always true only for thin lenses.

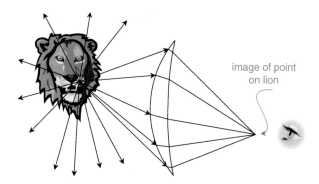

FIGURE 3.12 ▶ The object is a point on the lion's face, the lens forms an image of this point where multiple real light rays converge.

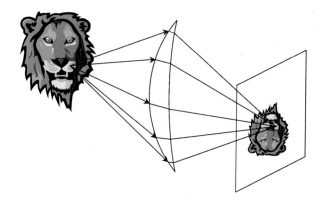

FIGURE 3.13 ▶ Each point on the lion's face can be imaged as in Figure 3.12, and the entire lion's face image is formed by the converging lens. Since the image is real, it can be captured by a paper screen, a photographic film, or a CCD (the detector in a digital camera).

the object, and for each point on the lion an image point is formed on the other side of the lens. If you put a screen at the image point, you see an *image* of the point on the lion. This is a real image of the object: There are real light rays converging at the image point. Therefore, if you put a piece of paper at the image point, the light rays will form a focused spot on the paper. In fact, you will see an image of the whole lion because, point by point, the rays traveling from the lion converge again on the other side of the lens to form a full image of the lion, as shown in Figure 3.13.

If there is no screen, you will see the image only if you stand behind the lens. From the side you will not see the image of the lion. The best way

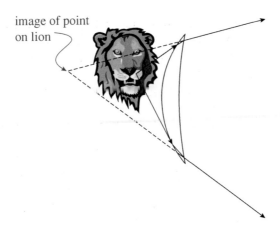

FIGURE 3.14 ▶ The object is very near the converging lens, and the lens is not strong enough to make the rays converge. The image is virtual.

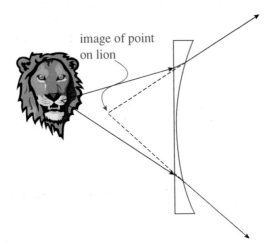

FIGURE 3.15 ▶ The diverging lens makes the rays diverge even more. The image is virtual.

to see the image is to place a screen where the image is formed, as shown in Figure 3.13, because the screen will direct some of the light into your eyes by, again, the process of scattering. The image is called *real* if (after passing through the lens) *the rays* from a given point of the object *cross each other, or converge, into a point*. A real image can be seen on a screen or on photographic film. To form a real image, we need a converging lens, but *converging lenses do not necessarily produce a real image,* as we shall see later.

The distance from the object to the lens is called the object distance, represented as o. The distance from the lens to the image is called the image distance, or i. It is important to always remember that the distance i may be where the rays from the object are focused, but this is *not* the focal length! Focal length refers to the place where **parallel incident** rays converge—this *requires* that the object be very far away! The image distance i is equal to the focal length f *if and only if* the object distance o is very large or at infinity (written as the symbol ∞).

If the object is very near the converging lens, the lens may not be strong enough to deflect the light rays coming from the object and make them converge into a point. Worse yet, we may have a diverging lens that spreads the rays even more than they were spreading before they got to the lens. In either case, after crossing the lens, the rays *seem* to come from a point at distance i behind the lens (with respect to your eye; see Figures 3.14 and 3.15). You can see the image with your eye if you look through the lens, but you cannot place a screen or a film where the image is and expect to see an image; no real rays cross in that location, or anywhere else for that matter!

This is a *virtual image*. We have already encountered virtual images: In a mirror, the image produced by reflection is also virtual. If you place a lightbulb near a mirror, there is no way to catch the image on a piece of white paper, no matter where you place it. To do this, you would need an additional lens (such as your eye or a camera taking a picture of the rays reflected by the mirror) to make the rays converge and form a real image. A *converging lens* forms a virtual image only when the object is closer to the lens than the focal length (Figure 3.14). A *diverging lens* always produces virtual images (Figure 3.15).

3.5 Three Easy Rays

Given a thin lens and an object, one can graphically draw rays to find the image size and position. With a diagram, you can see how the image is formed and where it is. This procedure is called *ray tracing,* and it is performed by very fast computers that trace many rays of light simultaneously, to simulate what happens in nature. Ray tracing is used in scientific applications, to produce sophisticated camera lenses, in animated movies, in computer graphics, and will be used by you in this course! It may seem complicated, but it is very simple to learn how to master ray tracing for a single thin lens. The trick is to focus on three particularly easy rays.

If we select a point on the object—for instance, the top of the lightbulb in Figure 3.16—we do not need to repeat the operation for each and every point of that same lightbulb. We only need to use three easy rays!

The *three easy rays* to be traced *for a converging lens* are shown in Figure 3.16 and described verbally hereafter:

- A ray from the object *through the center* of the thin lens. This ray goes *straight through* because at the center of the lens the two lens faces are

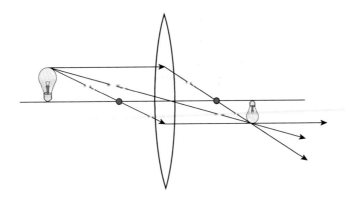

FIGURE 3.16 ▶ The three easy rays for a converging lens. The blue dots mark the distance f (focal length) from the lens. Remember that the distance is the same on either side of the lens. Note that for simplicity we do not draw the double refraction of each ray going through the lens, but consider only the trajectories of the rays going in and out of the lens. This is an acceptable simplification, since this and all other lenses considered here are very thin.

parallel to each other. There is actually a very slight shift for thicker lenses (see Figure 2.11), but it can be neglected for thin lenses like the ones we discuss here.

- A ray from the object to the lens, *parallel to the optical axis.* This ray goes *through the focal point on the other side of the lens*. The image point is where these two rays cross, so now we can draw the image as shown in the graph. It is an upside-down, or inverted, image.

- A ray from the object *through the focal point* on the left[2] of the lens. This ray will leave the lens traveling *parallel to the axis* and, of course, should go through the same image point already found. This is a good way to check the accuracy of the first two rays.

These are the three easy rays for a diverging lens, shown in Figure 3.17:

- As for the converging lens, the first easy ray goes straight *through the center of the lens*.

- The second ray is *parallel to the optical axis*. This ray is deflected out of the lens and appears *as if it came from the focal point on the left*[3]

[2]The third ray does not necessarily have to go through the focal point "on the left"; it has to go through the focal point on the same side of the lens as the object. Therefore, if the object is on the right, it goes through the right focal point.

[3]Again, this does not have to be the left side, just the same side as the object.

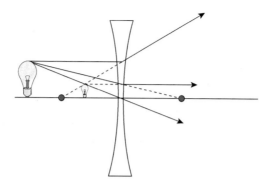

FIGURE 3.17 ▶ The three easy rays for a diverging lens. The blue dots again mark a distance f from the lens.

side of lens. The image point is where the first ray and the **extension** (dashed line) of the second ray cross.

- The third ray is the ray from the top of the object **heading in the direction of the focal point on the right of the lens**.[4] This ray will come out **parallel to the axis**, seeming to come from the point already identified by the first two rays. This ray may seem more complicated to draw. Just remember that light rays can always be inverted, and therefore, if this third ray were to come from the right traveling parallel to the optical axis, it will diverge on the left as if it came from the focal point on the right.

We see from these three easy rays that the image is virtual, upright, and smaller than the object. This is always the case for diverging lenses.

3.6 The Lens Formula

The lens formula (not to be confused with lens-marker's formula) can be used to quantitatively answer a variety of questions: Where is the image? Is the image real or virtual? How large is the image? The lens formula tells you how **object distance o** and **image distance i** are related to the focal **length f** of a lens:

$$\frac{1}{o} + \frac{1}{i} = \frac{1}{f}$$

---

[4]As for the previous notes, this is not necessarily on the right but has to be on the opposite side of the lens with respect to the object.

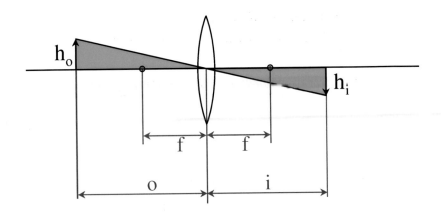

FIGURE 3.18 ▶   The magnification M can be easily calculated. The method is shown in this diagram. In the formulas, h_o is the height of the object and h_i is the height of the image. The formula is directly derived from the two similar triangles filled with green.

Figure 3.18 explains how to measure the magnification. Given a certain object distance o and a given focal length f, one can calculate the image distance i using the lens formula. The magnification M, then, is simply the ratio of i over o.

$$\frac{h_i}{h_o} = \frac{i}{o}, \text{ so the image size is } h_i = \frac{i}{o}h_o$$

$$\text{and the Magnification } M \text{ is } M = \frac{i}{o}$$

Be careful with the term *magnification*. Keep in mind that the magnification can be smaller than 1, and when this is the case, the lens really is *demagnifying* and the image produced is smaller than the object. The range of possible magnifications for a given converging lens varies dramatically, as the lens can be either magnifying or demagnifying, depending on how far the object is positioned from the lens.

The magnification is always a positive number: If the image distance i is negative (object and image are on the same side of the lens), take the modulus of i, or the absolute value—*ignore the minus sign*.

Let us now see some examples of how the lens formula and the magnification formula are used. For the next three examples, take a converging lens with a focal length of 10 cm. Notice that the focal length is always given by the manufacturer of the lens and cannot be changed (this is not true for zoom lenses, in which the focal length can be changed, but we are now talking about simple, single lenses).

Object distance greater than focal length

First, place the object 20 cm from the lens. Where is the image?

$$\frac{1}{i} = \frac{1}{f} - \frac{1}{o} = \frac{1}{10} - \frac{1}{20} = \frac{2}{20} - \frac{1}{20} = \frac{1}{20}$$

$i = 20\,\text{cm}$

So the image is 20 cm from the lens.

How big is the image?

The height of the object is h_o, and the height of the image is h_i.

$$\textbf{\textit{Magnification}} = \frac{h_i}{h_o} = \frac{i}{o}$$

In other words, the image size is to the object size as the image distance is to the object distance—the two lengths are proportional!

In this example $\dfrac{h_i}{h_o} = \dfrac{i}{o} = \dfrac{20}{20} = 1$ or $h_i = h_o$.

Therefore, for this example the image is the same size as the object.

Using the same lens, move the object to a distance of 50 cm away

$$o = 50\,\text{cm};\ f = 10\,\text{cm}$$

$$\frac{1}{i} = \frac{1}{10} - \frac{1}{50} = \frac{5}{50} - \frac{1}{50} = \frac{4}{50}$$

$$i = \frac{50}{4}\,\text{cm} = 12.5\,\text{cm}$$

Image size is now 5 cm, 1/4[= 12.5/50] of object size. Note that as we moved the object from 20 to 50 cm (further away from lens), the image got closer to the lens (from 20 cm to 12.5 cm).

There is a neat and easy way to understand (and more importantly to remember) this. Look at Figure 3.19. Light hitting point A of the lens is deflected as indicated by the angle labeled "deflection."

If the object is in the focal point (10 cm in Figure 3.19, left), the ray will be deflected by the lens in point A so that it comes out of the lens parallel to the optical axis. This deflection angle stays the same, no matter where the object is placed. So if, for example, we put the object at twice the focal length (20 cm; Figure 3.19, right), its image will be formed at a distance that keeps the deflection angle the same—that is, at exactly 20 cm. If we now move the object further away (as in Figure 3.20, left), the ray coming from the same

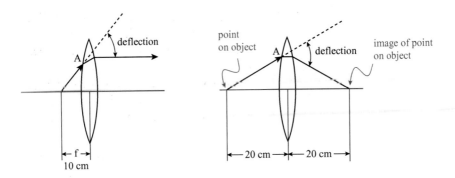

FIGURE 3.19 ▶ The deflection angle at point A on the lens is always the same, for any object distance.

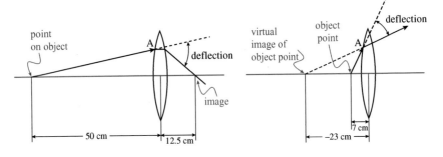

FIGURE 3.20 ▶ Same as in Figure 3.19, but with the object farther and closer.

point on the object and hitting the same point A will still be deflected by the same angle as before! The only difference is that the incoming ray hitting point A comes in at a lower angle (less steep) when the object is further away, causing the ray leaving the lens to be steeper and cross the axis earlier. Eventually, when the object is very far away, the ray will cross the axis at a distance f from the lens (which would be 10 cm in the earlier example). If we then move the object closer to the lens than the focal point—that is, 7 cm (Figure 3.20, right), we can still use the deflection angle to find the image position, even though in this case the image distance is on the same side of the lens as the object: The image is virtual!

Object distance smaller than the focal length

Using the same lens, move the object to a distance 7 cm from the lens, as shown in Figure 3.20, right.

$$\frac{1}{i} = \frac{1}{f} - \frac{1}{o} = \frac{1}{10} - \frac{1}{7} = \frac{7}{70} - \frac{10}{70} = \frac{-3}{70}$$

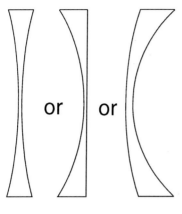

image distance $i = -70/3 = -23\,\text{cm}$ The minus sign indicates that the image is on the same side of the lens as the object, rather than on the opposite side as in examples 1 and 2. This *negative image distance* i indicates that the image is *virtual*. The size of the virtual image is 3.3 times the object size, since $M = \dfrac{i}{o} = \dfrac{23}{7} = 3.3$. For a converging lens, the image is virtual whenever the object distance is less than the focal length of the lens.

FIGURE 3.21 ▶ Diverging lenses.

Diverging lens

Now we are going to use a *diverging lens* with a focal length $f = -10\,\text{cm}$. This can have different curvatures, as shown in Figure 3.21, but it must be thinner in the middle than at the edges.

We must use $f = -10\,\text{cm}$ (negative focal length) in the lens formula.

$$\frac{1}{i} = \frac{1}{-10} - \frac{1}{o}$$

Note that the image distance i will always be negative (and the image will be virtual), no matter where we place the object—that is, for any value of the object distance o. For example, if $o = 10\,\text{cm}$

$$\frac{1}{i} = \frac{-1}{10} - \frac{1}{10} = \frac{-2}{10} = -\frac{1}{5}; \text{ therefore, } i = -5\,\text{cm}$$

The *virtual* image is 5 cm from lens, on the *same side of the lens* as the object. The image will be half the size of the object because $M = \dfrac{i}{o} = \dfrac{5}{10} = \dfrac{1}{2}$.

Remember to neglect any minus signs when calculating the magnification.

3.6.1 Note on Magnification

As previously mentioned, the magnification is always lower than 1 for diverging lenses. In other words, diverging lenses are always demagnifying lenses. On the other hand, the magnification can vary dramatically for converging lenses, which can be magnifying or demagnifying, depending on where the object is positioned with respect to the lens. See Figure 3.22 for various

When the object (man with umbrella) is further than the focal length (blue dots), the image is real, upside down, and smaller than the object.

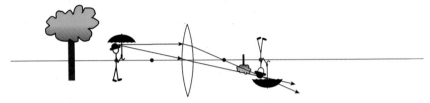

When the object is at twice the focal length (magenta dots) from the lens, the magnification is 1.

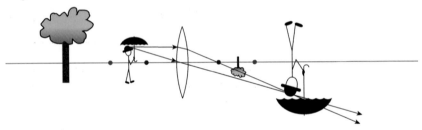

When the object is between the focal length and twice the focal length (between magenta and blue dots), the magnification is greater than 1.

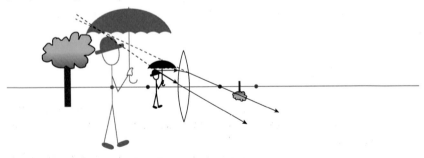

When the object is closer to the lens than the focal length, the image is virtual (gray man), upright, and the magnification is greater than 1.

FIGURE 3.22 ▶ The magnification increases as the object distance decreases. (Diagrams adapted from Eugene Hecht, *Optics*, Addison-Wesley Publishing Company, Reading, MA, 1997)

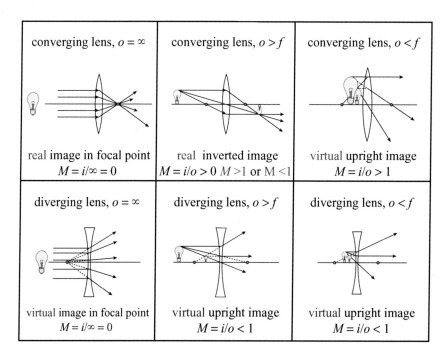

converging lens, $o = \infty$	converging lens, $o > f$	converging lens, $o < f$
real image in focal point $M = i/\infty = 0$	real inverted image $M = i/o > 0$ $M > 1$ or $M < 1$	virtual upright image $M = i/o > 1$
diverging lens, $o = \infty$	diverging lens, $o > f$	diverging lens, $o < f$
virtual image in focal point $M = i/\infty = 0$	virtual upright image $M = i/o < 1$	virtual upright image $M = i/o < 1$

FIGURE 3.23 ▶ Summary of the images and magnifications resulting from all possible cases for converging and diverging lenses: object at infinity, object farther than the focal length, and object closer than the focal length.

object positions. In these diagrams, notice that one object (the man with the umbrella) moves, while another object (the tree) is stationary. Compare the relationship between the two objects on each side of the lens for each diagram. In Figure 3.23 we summarize all possible images and magnifications produced by converging and diverging lenses.

3.7 Lens Aberrations

The formulas and ray tracing method we have discussed so far work accurately as long as the lenses are thin and, more importantly, as long as we are in the *paraxial approximation*, meaning that we only consider *rays of light close to the optical axis*. Selection of only paraxial rays can be done either using a lens with a very small diameter or using a larger lens with an iris aperture that limits the lens diameter actually used. Decreasing the effective diameter—or closing the aperture—limits the brightness of the image—that is, the *luminosity*. Consider an object and a lens. Rays from the object hit all points on the surface of the lens, not just the points hit by the three easy rays we considered before. Each and every portion of the lens is producing the image of every point of the object. You may therefore use a small lens to take a picture of a large tree, or you may cover a portion of the lens and still obtain the same image! If, for example, you cover half of the lens, the other half (which now has a D shape), you will obtain the exact same image of the object, with the same magnification and position as if the entire lens was uncovered. This image, however, will have half the intensity. The reason for using a larger lens is to increase the lens's luminosity. That is why the best telescopes—those that can see the faintest stars—have very large diameters. The same applies to camera lenses: The best cameras have lenses with the largest diameters, allowing them to take photographs in dim light conditions by opening the aperture all the way and using the entire diameter of the lens.

Departing from the paraxial condition—for instance, when taking a photograph with the aperture wide open—introduces problems in image clarity and shape. These problems are called lens aberrations. There are different kinds of aberrations. Chromatic aberrations arise from the fact that the index of refraction n varies for different wavelengths of light (colors). Monochromatic aberrations, on the other hand, occur even if a single wavelength of light is used. Monochromatic aberrations include spherical aberration, coma, astigmatism, field curvature, and distortion. Here we will only focus on the types of aberrations that mostly affect photography.

3.7.1 Chromatic Aberrations

As described in Chapter 2, the index of refraction of glass is a little different for each different wavelength of light, a phenomenon known as dispersion. Rays of different colors are separated by the lens, as blue light rays are deflected more than red light rays. Consequently, the focal length of a simple lens for blue light is different from the focal length of the same lens for red light, as shown in Figure 3.24. The difference is not very large, but it is enough to cause trouble. For example, you cannot capture both red and blue polka dots of a woman's skirt with a good, sharp focus at the same time if you are using a camera with a simple lens. If you focus on the red dots, the blue dots will become a little fuzzy and vice versa. If you can see the image of the red dots as perfectly in focus, the image of the blue dots will be in focus at about 0.2 mm in front of the film. By the time the rays get to the film, they will have already re-diverged a little, producing a slightly blurry image of the blue dots. The blur is worse if the lens aperture is large because the rays diverge more over the 0.2 mm distance. This is because rays of light farther from the axis of the lens must be deflected more (at a sharper angle) to be brought into focus. The more rays of light are deflected by the lens, the greater the effect of chromatic aberration. If you don't need a large lens aperture, however (for example, if you only want to take pictures outside on sunny days), there is a simple way around this problem-just use a small lens aperture.

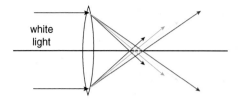

FIGURE 3.24 ▶ Blue light is deflected by the lens more than red light, so the focal point for blue is closer to the lens than the focal point for red. Blue and red objects cannot be in focus at the same time. Closing an iris aperture—that is, selecting the central or paraxial rays only—reduces chromatic aberration.

Most cameras have achromatic lenses that produce sharper pictures and that can be used in dimmer light. Achromatic lenses are built by combining a strongly converging lens, made of low-dispersion glass, with a weakly diverging lens of high-dispersion glass. With this combination, the dispersion cancels out, while net focusing remains. You might wonder why lens makers don't simply use a type of glass with no dispersion. There is no such glass or plastic! Another common question is, Why not cancel the dispersion by making half of the lens with glass of the opposite characteristic? Again, this is impossible since all transparent materials deflect blue more than red.

3.7.2 Spherical Aberration

This type of aberration is color-independent and arises from the fact that rays hitting the outer diameters of a lens are deflected more than paraxial rays, the rays hitting the lens inner diameters. This is true for both converging and diverging lenses. The net result does cancel if you use a combination of diverging and converging lenses, as shown in Figure 3.25. Another way to eliminate spherical aberration is to use a small aperture, but, again, this can only be done in conditions of bright light.

Chromatic and spherical aberration, as well as coma and astigmatism, deteriorate the image clarity. Field curvature and distortion, on the other hand, deteriorate the image shape. In other words, they deform the image.

Distortion: It is not enough that a picture is in sharp focus for all colors and for all rays passing through the lens. Suppose you take a picture of a rectangular building with a rectangular grid of windows. These should produce an image on the film of rectangles, not slightly curved lines. However, there are two types of common distortions that cause such curved images: the pincushion distortion and the barrel distortion, shown in Figure 3.26.

These distortions arise from the fact that the image magnification varies with the image radius. With pincushion distortion, the center of the image has lower magnification than the edges, while with barrel distortion, the center has higher magnification than the edges.

The design of a good-quality camera lens is complicated (and therefore the lens is very expensive) because it has to compensate for all the possible aberrations for all possible object distances, including objects near the center of the field of view and near the edges. The solution to the problem lies in making compound lenses. This is done by stacking different lenses, each lens curved and spaced from the next lens in the proper, particular way. Computer ray tracing programs are used to do trial-and-error simulations of compound lenses to find the combination of lenses with the best performances, thereby minimizing

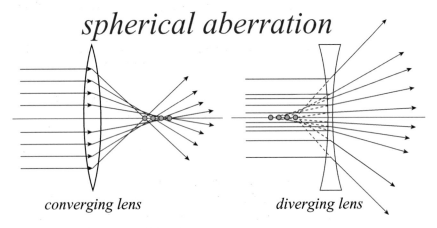

spherical aberration

converging lens   *diverging lens*

FIGURE 3.25 ▶ The converging lens moves the focus to the left when using outer diameters of the lens. The diverging lens moves the focus to the right when using the outer diameters. Since these two lenses compensate for each other, in camera lenses a combination of diverging and converging lenses is used to minimize spherical aberration. Unlike those in Figure 3.9, these must have different curvatures and focal lengths.

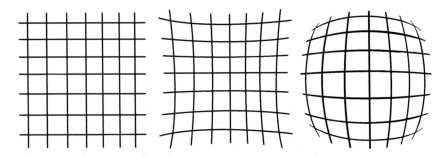

FIGURE 3.26 ▶ On the left is the object, at the center its pincushion distorted image, and on the right is its barrel distorted image.

distortions and spherical aberration. Only then are the lenses physically mounted in stacks. For most purposes, it is not worthwhile to buy the absolute best lens available, since the imperfections for even an average-quality lens are only very slight. Even cheap lenses can take pictures of good quality as long as the lens aperture is kept small. However, if your interest is to take high-quality or professional photos when you really can't use a tripod or a flash, such as early-morning wildlife photography, you must use larger lenses—for example, a lens with a very large diameter (e.g., 300 mm) that increases luminosity and allows for a fast exposure (which you need with no tripod and no flash). Such lenses are always compound, and you may have to spend more than $10,000 on a single lens! Not only is it expensive, but it is heavy, sometimes weighing 20 pounds! Imagine carrying that around the woods on long hikes. The life of a photographer is not as easy as you might think!

Coated lenses: All high-quality camera lenses have coatings to reduce the amount of light intensity that is lost to reflection from the lens surfaces. These are non-reflective coatings. Some lens coatings may have a noticeable magenta tinge. This is because a small amount of blue and red light is reflected and makes the coating appear magenta. However, the best-quality coatings are perfectly colorless, and they do not allow for any light to be reflected. Let us see in detail why such non-reflective coatings are necessary to produce a good photograph.

Not all the light incident on the surface of the glass lens penetrates the lens and gets refracted. Part of the light is reflected back. The loss of light intensity on the camera film from this reflection is quite serious because, as more than one lens is usually used to avoid aberrations, the light goes through many surfaces and the amount of light intensity lost at each surface adds up.

Example: If a compound lens is composed of 6 lenses, there are actually 12 surfaces.
Even if only 8% of the light intensity is lost to reflection on each individual surface,

$$0.92^{12} = 0.367$$

after traveling through all 6 lenses, only 36.7% of the original light reaches the film!

Without a nonreflective coating, you would really be better off using a simple lens and avoiding aberrations instead—and simply use a small aperture!

This is why, especially for compound lenses, a non-reflective coating is always used. The coating consists of a thin film that has just the right index of refraction and just the right thickness so that reflected light from the front and back surface of the coating cancel by destructive interference. The light cannot be reflected and has to go through the lens.

3.8 Questions

1. Which lenses are converging? Which are diverging?

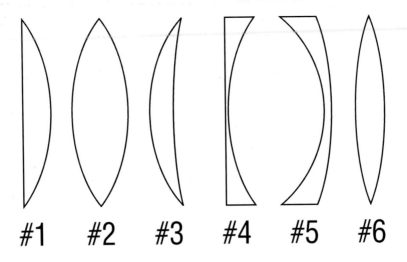

#1   #2   #3   #4   #5   #6

2. Which converging lens has the shortest focal length?
3. What does one mean by the "focal length" for a converging lens? For a diverging lens? When is the focal length a positive number? When is it negative?
4. What does one mean by a strong lens?
5. Find the position and size of the image by using:

$$\frac{1}{i} + \frac{1}{o} = \frac{1}{f} \quad \text{and} \quad \frac{i}{o} = M$$

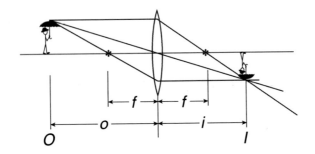

The object O is at a distance o from the lens, while the image I is at a distance i from the lens. If the focal length is $f = 25$ cm, $o = 100$ cm, and height $h_o = 2$ cm, where is the image? And how high is the image? ($i = ?$ and $h_i = ?$)

6. A lens has a focal length of 20 cm. You place an object 100 cm from the lens; where is the image? (Where should you put a screen or a film to obtain an in-focus image?) How large is the image compared to the object?

7. If you place the object 10 cm from the lens in Problem 6, where is the image?

8. A lens has a focal length of 5 cm, and the object is 50 cm away from the lens. What is the image distance?

9. The following lens has a focal length $f = 3$ cm. (Foci are marked *.) Draw the "easy rays" and find the position of the image graphically, and then use the lens formula to verify your results for the two object distances shown. What is the magnification in each case? Copy the drawing carefully for your answer.

(a)

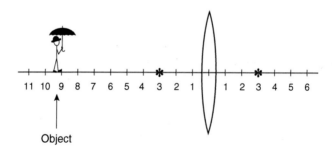

Object

(b)

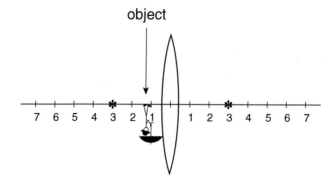

object

10. In the following diagram, find the position of the image by tracing the three easy rays. The focal points of lenses are marked *.

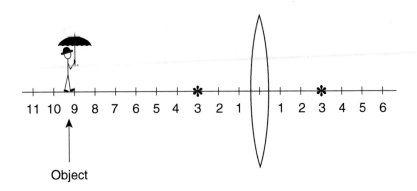

11. Show the three "easy rays" from the top of the object through the lens. The symbol * marks the focal length on either side of the lens. Show the image.

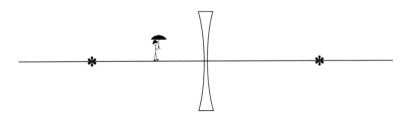

12. Show the three "easy rays" from top of the object through the lens. The symbol * marks the focal length on either side of the lens. Show the image.

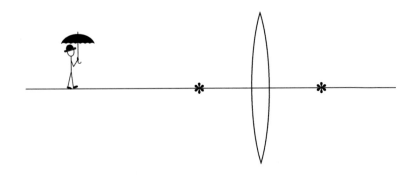

13. You are lying on the grass with a friend. A daffodil 0.2 m high and 1 meter away seems as big as the oak tree that is 100 m away. You see them as shown in the following figure. What is the height of the oak tree?

Chapter 4

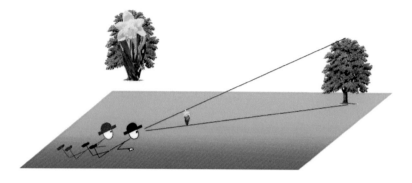

The Eye

The lenses discussed in Chapter 3 are useful in understanding how the eye works, its defects, and the corrections that appropriate lenses can provide. The text in this chapter was adapted from Eugene Hecht's book, *Optics*, which we recommend for a more in-depth discussion.

The human eye is a complex organ composed of several parts, arranged as shown in Figure 4.1.

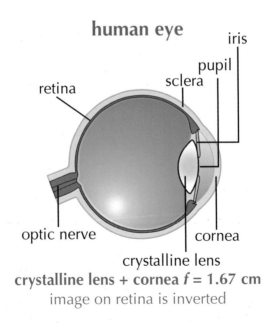

human eye

crystalline lens + cornea *f* = 1.67 cm
image on retina is inverted

FIGURE 4.1 ▶ The parts composing a human eye.

Physics in the Arts, Revised Edition
© 2012 Elsevier Inc. All rights reserved.

Rays of light cross a system of two lenses and then converge, in focus, on the retina to form a real inverted image. The image has to be real—that is, actual rays must converge on the retina. If the image is not real and not formed by actual intersecting rays, the cone and rod cells in the retina will not be able to detect the image, convert it into a train of pulses, and transmit it to the brain via the optic nerve. The left and right eyes produce different upside-down images. The two images are then blended together and inverted again by the brain. The two images seen by the two eyes are identical to each other when the objects are very far away, but their difference increases as objects become closer. We do not realize that our two eyes see two different images until we close them alternately and observe the right and the left image separately. The difference in the images is important because it makes it possible for our brain to observe the three-dimensionality of nearby objects and to provide an estimate of their distance.

The eye is contained within a resistant yet flexible eyeball, the sclera, which contains jelly-like fluids. Except for the front area, called the cornea, the sclera is white and opaque. The cornea is the curved surface bulging out from the eyeball. The cornea is slightly flattened at the center to reduce spherical aberration. It is a hard, transparent, convex layer of cells, forming a converging lens. It is the stronger of the two lenses in the eye, so it has a lower focal length and a higher curvature than the other lens. Most of the deflection of light rays by the cornea takes place at the air-cornea interface because the refraction index of the cornea ($n_{cornea} = 1.376$) is much greater than that of air ($n_{air} = 1$). It is similar to that of water ($n_{water} = 1.33$), which is why we can't see well underwater: Rays are not deflected much at the water-cornea interface and reach the retina out of focus. After the cornea is the iris, which is immersed in a watery fluid and serves as a lens aperture. Just like the aperture stop in a camera, the iris controls the amount of light entering through the pupil, the hole allowing light to enter the eye. The iris is made of circular and radial muscles so it can expand and contract, so the diameter of the pupil varies between 2 mm in bright light and 8 mm in darkness.

Immediately behind the pupil of the iris is *the crystalline lens*. This lens has the shape of a small lentil bean or an M&M candy, with a diameter of 9 mm and a thickness at the center of about 4 mm. The crystalline lens has a very complex structure, with 22,000 concentric layers (as in an onion), with indices of refraction that vary slightly from one layer to the next. At the dense, inner core, $n = 1.406$, while at the surface $n = 1.386$. A manmade lens with these characteristics is called a GRIN lens, from GRadient INdex.

The crystalline lens provides fine focusing through changes in its shape. Behind the crystalline lens is a jelly-like fluid called vitreous humor, which differs from the fluid in front of the crystalline. The vitreous humor contains microscopic particles, which are dead cell debris floating around within it. You can see these floaters by looking at the sky on a bright day. They appear as transparent, amoeba-like objects floating across the field of view.

At the back inner surface of the eyeball is the retina. This is a thin, semi-transparent layer of light receptor cells. The human eye has two kinds of light sensitive cells: rod cells and cone cells. The rods provide low-resolution vision and function well in dim light, while the cones provide high-resolution color vision but only function well in bright light. There is a small depression of about 2.5–3 mm in diameter at the center of the retina that is known as the yellow spot, or the macula. This spot contains only cone cells and no rod cells. At the center of the macula is the fovea, with a diameter of 0.3 mm. To give you some perspective, keep in mind that the image of a full moon on the retina is 0.2 mm. In the fovea the cones are thinner (with a diameter 1.5–3 μm) and more densely packed than anywhere else in the retina. Since the fovea provides the highest resolution and most detailed visual information at the center of our field of view, we continuously move our eyeballs so that light coming from the object of primary interest always falls on the foveas.

Each cone cell has its own connection with an individual nerve fiber, while approximately 100 rod cells connect to a single nerve fiber. All nerve fibers come together in the optic nerve, which brings the signal from the retina to the brain. The optic nerve connects with all the nerve fibers converging from all over the retina. In the point in which the optic nerve is attached to the retina, there is a region in which we do not see. This is called the blind spot.

Look at Figure 4.2, to localize your blind spots.

X   **1**   **2**

FIGURE 4.2 ▶   To localize the blind spot of your right eye stand at a distance of about 10 inches from this figure, close your left eye, and focus your right eye on the X. You will only be able to see the 1 because the 2 falls on the blind spot. If you now move closer, the 2 will reappear and the 1 will vanish. You can test this on your left eye as well, but you have to rotate the book upside down.

4.1   Accommodation

As just mentioned, the crystalline lens provides fine focusing of the image from the cornea and the pupil. The focal distance of the system of the two

accommodation of the eye

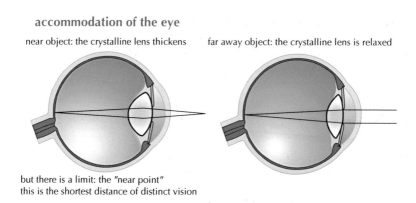

FIGURE 4.3 ▶ Accommodation of the eye.

lenses in the eye, the cornea and the crystalline together, is 1.67 cm. If an object is at infinity, the image is exactly in the focal point, which is on the retina. If the object comes closer, the image will be formed in focus *after* the retina, and the image on the retina will be fuzzy. The crystalline lens, which is flexible, can compensate for this effect. It becomes thicker in the middle, becoming a stronger lens with a smaller radius of curvature and a shorter focal distance. After the crystalline lens thickens, the images will be perfectly in focus on the retina. This process of thickening is called ***accommodation***, and is shown in Figure 4.3.

As the object moves closer to the eye, the crystalline lens accommodates more and more until it reaches a maximum point beyond which it cannot thicken anymore. That maximum crystalline thickness defines *the near point* of the eye. This is the shortest distance of distinct vision. You can easily find your own near point. With one hand, cover one eye, extend your other arm, and look at your finger. Now, bend your arm and slowly bring that finger closer to the eye, keeping it in focus. As you do this, your crystalline lens is thickening! As the finger approaches the eye, it will get closer and closer to the near point. When it passes the near point, you will not be able to focus on the finger anymore.

The near point-the shortest distance at which you can focus-varies from person to person, and it also changes for the same person with age. As a person gets older, the crystalline lens loses elasticity and the near point is at greater distance away from the eye. With a normal eye, the near point is 7 cm for a teenager, 12 cm for a young adult, 28–40 cm for a middle-aged person, and about 100 cm for a 60-year-old. That is why the normal eye after a certain age needs reading glasses. These move the near point closer. As you may have imagined already, this correction is done using glasses with converging lenses. The myopic eye may never need reading glasses.

All mammals accommodate their vision this way. Fish, on the other hand, can only move the lens in their eye away or toward the retina to adjust the focus, just as we do when using a camera. Some mollusks elongate or contract the eyeball depth, and birds of prey accommodate by changing the curvature of the cornea.

4.2   Eyeglasses

Macfarlane, an anthropologist, and Martin, a historian of glass instruments, both from Cambridge, United Kingdom, speculate that the development of glass was fundamental for science itself. Lenses are the major components of telescopes for looking at the stars and of microscopes for looking at cells. Imagine life without glass before the first synthesis of transparent acrylics and plexiglass. There would have been no windows to keep the warm air inside while letting in sunlight, so the time for reading and studying would have been far shorter, slowing down progress considerably. There would have been no bottles to foster the development of alchemy and chemistry. Ancient cultures that drank tea instead of wine, such as the Chinese and Japanese, never had a good reason to develop glass-blowing skills because ceramic pottery had far superior characteristics to hold hot liquids without cracking. Macfarlane and Martin infer that the stagnation of those cultures in and after the Middle Ages was due to the absence of glass. However, these are only speculations, and there is of course no proof of how the development of glass technology would have affected those cultures.

Eyeglasses were invented first in Florence, Italy, in 1299. These were convex, converging lenses only. Glasses using diverging lenses did not appear until about 150 years later. Negative, diverging lenses were actually first produced in 1267, but they were not used in glasses for about 200 years. Interestingly, it was considered improper in Europe to use eyeglasses in public until the 1800s, and, therefore, we have very few paintings of people wearing glasses.

A *normal eye* is an eye capable of focusing parallel rays on the retina while in relaxed condition.

The *dioptric power* $[DP = 1/f(m)]$ of the unaccommodated normal eye is 60 D (1 diopter $= 1\,D = 1m^{-1}$), of which 43 D is from the cornea and 17 D is from the crystalline lens. The measure 60 D corresponds to a combined focal length of the two lenses of 1.67 cm ($1/60 \times 100\,cm$).

The *far point* is the maximum distance of a point whose image lies on the retina, or the maximum distance of distinct vision. For a normal eye, the farthest point focused on the retina, the far point, is located at infinity. Infinity, for all practical purposes, is anywhere beyond about 5 m from the eye's lens. When the focal point of the eye's system of two lenses does not lie on the retina, the eye is nearsighted, farsighted, or astigmatic. This may be a result of abnormal changes in the lenses (refractive index or curvature) or the dimensions of the eyeball, which alter the distance between the lens system and the retina. The second problem, abnormal dimensions, is the most common cause of imperfect vision.

4.3 Nearsighted Eye

A deep eyeball causes the eye to be *nearsighted*, or *myopic*, as shown in Figure 4.4. This can be corrected using a negative lens. In a nearsighted eye, parallel rays converge before the retina and have diverged again by the time they reach the retina. Therefore, faraway objects appear blurry and larger than they should. The far point is closer than 5 m—for instance, at 2 m. To correct for this problem, we have to add a lens to the optical system. Only when the three lenses are combined (the eyeglass lens, the cornea, and the crystalline lens) can we produce an image of parallel rays in focus on the retina. Since the nearsighted eye cannot focus on faraway objects, we add a *diverging lens* that forms a virtual image of distant objects closer to the eye. For objects at infinity, the lens creates a virtual image located at the far point (2 m). No accommodation by the crystalline lens is required. If the object is closer than infinity, the virtual image is closer than the far point—say, 1 m away in the preceding example. This image is then viewed by the cornea and crystalline lenses, which cast a real, focused image on the retina, but only after accommodation. Another result of using these diverging lenses is that they move the near point a little further. This is why nearsighted people prefer to take their spectacles off when performing precision tasks, such as threading needles or

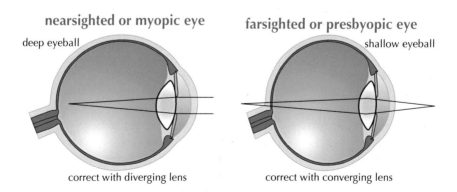

FIGURE 4.4 ▶ Nearsighted and farsighted human eyes.

reading small print. Without the diverging lens, they can then bring the object closer to the eye and increase the magnification.

4.4   Farsighted Eye

A shallow eyeball causes the eye to be *farsighted*, or *hypermetropic*, also shown in Figure 4.4. Objects at infinity are seen clearly, but nearby objects are blurred because the rays converge after the retina. Recall that the crystalline lens accommodates only up to a point, reaching its limit at the near point. The near point is further than norm for a farsighted eye. This defect can be corrected with a *converging lens*, which makes the rays converge earlier, on the retina. Such lenses effectively move a close object further away, beyond the near point, where the eye can see clearly. The lens will form a distant, virtual image that can be clearly seen by the cornea-crystalline lens.

Recall that diverging lenses are always demagnifying, while converging lenses with nearby objects are always magnifying. You can, therefore, always tell when looking at a person in front of you wearing eyeglasses which kind of lenses they are using. In this case the person's eyes are the objects, then the lenses are their eyeglasses, and the image is the image of their eyes, which you can see standing in front of them. If their eyes appear smaller through their glasses the lenses are diverging, so you know the person is nearsighted. If their eyes are magnified by their glasses, you know the lenses are converging and they must be farsighted.

For the same reason, nearsighted people always see the world a lot smaller when they wear their glasses. This is not only due to the fact that sharp objects are smaller than blurry ones. It is also due to the demagnifying effect of all diverging lenses.

4.5   Astigmatic Eye

Astigmatism is the most common eye defect. Almost everyone has some degree of astigmatism, but in most cases this is so slight that people don't even know they have it. It arises from a nonspherical curvature of the cornea. In particular, the curvature along different meridians differs, as in an eggshell. Most frequently the curvature of the cornea along a meridian is greater than the curvature in the plane perpendicular to that meridian. The result is a blurred and/or distorted image on the retina. This defect can often be corrected using *cylindrical lenses*, first introduced in 1825 but only commonly used after 1862.

Chapter 5

Photography

5.1 The Camera

The oldest preserved black–and–white photograph was taken by Joseph Nicéphore Niépce in 1826. It was produced by a dark box, a lens, and a light from his window view projected onto a pewter plate that was coated with bitumen, a petroleum derivative. After a day-long exposure, the plate was washed with a mixture of oil of lavender and white petroleum, which dissolved away the parts of the bitumen that had not been hardened by light. The result was the permanent direct positive photograph on pewter, rendering a view of the outbuildings, courtyard, trees, and landscape as seen from that window, which can still be seen today at the University of Texas at Austin. Soon after this first photograph was taken, Louis Daguerre (from whom the name *daguerreotype* for photographic prints on glass is derived) and William Henry Fox Talbot invented the photographic process in which a *negative* was exposed in the camera, developed in a dark room, and then printed on sensitive glass or paper. Their cameras were conceptually similar to modern ones and are the easiest of the optical instruments to understand. The key parts that form any camera, be it the simplest ancient version or a very sophisticated modern one, are shown in the schematic diagram of Figure 5.1:

- *The lens*, *L*, produces a *real image* of the object that one wishes to capture on the film or CCD. It is usually mounted on a barrel cap that can

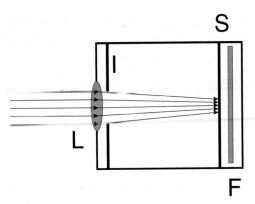

FIGURE 5.1 ► Schematic representation of a camera.

be moved in and out, allowing the photographer to adjust the distance between the film and the lens in order to focus the image.

- ***The shutter, S***, is normally closed but opens for a small interval of time to allow light from the object to enter the camera. The shutter may be placed near the lens ("lens shutter") or near the film ("focal plane shutter").

- ***The iris aperture, I***, also called an *f-stop*, is an adjustable diameter aperture (hole or orifice) located near the lens. This controls the amount of light that enters the camera.

- ***The film, F***, consists of a base, a polymer sheet, or a glass plate that is coated with a light-sensitive *emulsion*. This emulsion contains grains of silver salts (silver halides, such as AgBr) in a gelatin.

- When light strikes the emulsion, the emulsion grains are modified. Chemical treatment of the emulsion, including *developing* and *fixing*, removes the non-modified emulsion grains and transforms the exposed grains into metallic silver, which is black.

Let us now look in more detail at the operation of the various parts of the camera.

5.2 Focusing the Camera

Focusing the camera involves adjusting the distance between the lens and the film so that the *real image* of the object being photographed is located on the film plane, not in front or behind it. Why is that important? Consider a point object being photographed. If the camera is not focused properly, the light rays will not converge to a point on the film. Rather, they will produce a fuzzy spot of a larger size, and the picture will be blurred.

For an example, consider a simple *35 mm camera*. This label refers to the size of the film, which is 35 mm wide and 24 mm high. The focal length of the lens is most commonly $f = 50$ mm. If you wish to photograph a faraway mountain, the object distance is very, very large. When the object distance is more than 100 times greater than the focal length of the lens (5 m in this case), photographers refer to the object as being "at infinity." They use the standard mathematical symbol for infinity to represent this distance: ∞.

When the object is very faraway, we write the lens formula as follows:

$$\frac{1}{f} = \frac{1}{i} + \frac{1}{o} \qquad o = \infty; \qquad \text{then} \qquad \frac{1}{o} = \frac{1}{\infty} = 0 \qquad \text{and} \qquad i = f$$

In other words, if o is very large, $\dfrac{1}{o}$ is very small—so small that it can be neglected! The image distance is then equal to the focal distance. Therefore, when using a lens with a focal distance $f = 50$ mm, both the film and the image must be at 50 mm from the lens in order to take a photograph of a mountain.

Suppose now that you want to take a closeup picture of a flower about 20 inches away.

$$20" \approx 0.5\,\text{m} \qquad o = 0.5\,\text{m} = 500\,\text{mm}$$
$$f = 0.05\,\text{m} = 50\,\text{mm}$$

Using the lens formula again:

$$\frac{1}{f} = \frac{1}{i} + \frac{1}{o} \qquad \frac{1}{i} = \frac{1}{f} - \frac{1}{o} = \frac{1}{50} - \frac{1}{500} = \frac{10-1}{500} = \frac{9}{500} = 0.018 \text{ and}$$
$$i = 55.55 \approx 56\,\text{mm}$$

The distance between the lens and the film must be 56 mm. As you can see, a very small motion of the lens—in this case, $56 - 50$ mm $= 6$ mm (or about 1/4 inch)—is all that is needed to change focus from infinity to 0.5 m (2 feet). Thanks to the fact that only a small variation in image distance is necessary to focus objects at any distance, cameras can be small and portable!

The change in image distance with focusing is small but by no means negligible. Light rays diverge a lot in a small distance, so accurate focusing must be done, or the image will be blurry. The shutter is placed either near the lens or near the film. The *focal plane shutter* is a double curtain mechanism that lies directly in front of the film—that is, as close as possible to the focal plane. Initially, both curtains cover the film (frame 2 in Figure 5.2). When you press the button to take a photograph, the first curtain starts to move across the whole film width, exposing the film to the light entering from the lens (frames 2–7). After a certain time (which the photographer sets as the *exposure time*), the second curtain starts to move in the same direction as the first curtain, covering the film again and cutting off light exposure (frames 8–12). For short exposures, the two curtains move very closely together, keeping a constant distance between them. Each point in the film is exposed for the exact same amount of time (5 time-units in Figure 5.2). In real cameras, the curtains move continuously, not in steps as in the schematic representation of Figure 5.2. The sequence reported is only a simplification to aid your understanding.

All cameras have a set of numbers that indicate the *exposure time*, or the time that each point on the film is exposed to light as the shutter curtains move across the front of the film. The exposure times are usually $1000, 500, 250, 125, 60, 30, 15, 8, 4, 2, 1, B, T$. This actually means 1/1000th of a second, 1/500th, 1/250th, etc., down to 2, meaning 1/2 of a second, and 1, meaning simply 1 second. These are the times a photographer has at his disposal. Notice how they form a sequence varying by factors of 2, approximately.

Exposure times vary by factors of 2. For each step toward the left, the exposure time doubles, and to the right it is cut in half. This is not perfectly accurate, as you can see by comparison with the geometric progression in the following chart, but it is very close. We will soon see how this is very useful when selecting the appropriate settings in order to capture the desired photograph.

When a camera is set on B, which stands for "bulb," the shutter is opened when the button is pressed and closes again when the

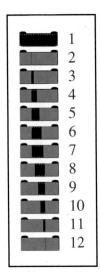

FIGURE 5.2 ▶ When taking a photograph the curtains move in front of the film so that each point of the film is exposed for the same amount of time, thus each point receives the same illumination.

Exposure Time in Seconds	1	1/2	1/4	1/8	1/15	1/30	1/60	1/125	1/250	1/500	1/1000
Geometric Progression	1	1/2	1/4	1/8	1/16	1/32	1/64	1/128	1/256	1/512	1/1024

button is released. When the camera is set on T, which stands for "time," the shutter is opened by pressing the button and closes only by pressing it again.

5.3 Choosing the Exposure Time

If the object to photograph is moving, it may appear blurred, depending on how much it moved while the shutter was open. The distance that an object moved is equivalent to the speed × time. In other words, $d = v \times t$. A track star running a 4-minute mile travels at a speed of 7 m/sec. If you set your camera's exposure time to 1/100 sec, the runner will have moved about 3 inches during the exposure time, and the photograph will look quite blurry. This reveals that the exposure time of 1/100 sec is too long (or too slow) to take a good action shot of the runner.

$$7 \text{ m/sec} \times 1/100 \text{ sec} = 0.07 \text{ m} = 7 \text{ cm}$$

A person walking at 1 m/sec moves only 1 cm in 1/100 sec. It is therefore acceptable to take photographs using an exposure time of 1/100 sec with people moving at this speed.

Even when the object is stationary, if the camera is held by hand, the maximum exposure time that can be used is about 1/60 sec. This is because of the natural and inevitable movements of the human body. For any exposure time longer than 1/60 sec, as necessary for taking photographs at dusk, special attention is required to not move the camera. If the camera moves, the image of the stationary object moves on the film, and the image will appear blurred. At 1/60 sec there is not a blurring problem for any handheld camera. Even at 1/30 sec, if you hold your breath and lean against the wall, you may be able to take a good picture. However, for longer exposure times, such as 1/15 sec or 1/2 sec, the camera must be held by a tripod or set on a table or car roof.

5.4   Choosing the Aperture

The iris aperture is a hole, or orifice, located immediately after the lens. It is made of many thin metal blades that can be pivoted along the rim in order to adjust the diameter of the opening. Cameras normally do not show the actual diameter of the aperture. What is shown, printed on the lens mounting, are the various *f numbers* that are mathematically represented by the symbol *f/*. The diameter of the iris aperture is calculated by

$$f/ = \frac{f}{Diameter}$$

where f is the focal length of the lens and Diameter is the diameter of the aperture—the effective diameter of the lens portion being used. For example, if a camera has a lens with 50 mm focal length and the aperture is set at an f number of 4 (this is written *f/4, not f/ = 4*), then the diameter is actually 50 mm/4 = 12.5 mm.

The sequence of *f numbers,* or *f/*, that normally appears on the aperture adjustment control is

f/	1	1.4	2	2.8	4	5.6	8	11	16	22	32	45

But what happens when you change the f number by two steps, from $f/4$ to $f/8$—that is, from 4 to 5.6 (one step) and again from 5.6 to 8 (another step)? A photographer calls this change "two stops" because *aperture stop* is the colloquial, commonly used name for the iris aperture. The diameter of the aperture is changed in this example from 50/4 = 12.5 mm to 50/8 = 6.25 mm.

Through comparing these two diameters, it is apparent that changing the aperture by two steps changes the diameter by a factor of 2. The *area* of the aperture changes by a factor of 4 (because the area of a circle is proportional to the radius squared). Changing the aperture by one step changes the diameter by a factor of 1.4:

$$\frac{2.8}{2} = 1.4; \quad \frac{4}{2.8} = 1.4; \quad \frac{5.6}{4} = 1.4; \quad \text{and so on.}$$

One step changes the area by a factor of 2:

$$1.4 \times 1.4 = 2$$

Each single step in f/changes the lens area twofold and therefore also changes the amount of light on the film by a factor of 2.

Besides its primary function of controlling the amount of light entering the camera, the value of the f number also affects another aspect of the picture: *the depth of field*, defined in Figure 5.3.

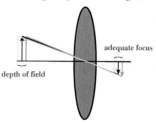

Depth of field
the range of object distances that produce adequately focused images on film

adequate focus

depth of field

FIGURE 5.3 ▶ Definition of depth of field.

5.5 Depth of Field

Suppose you are taking a photograph of an object that is 4 m away from the camera. If you want to take a good picture, you will naturally want to adjust the lens so that the object is in perfect focus. Objects farther away and in the background—say, 5, 7, 8, or more meters from the main object—will be out of focus. These background objects will become increasingly more and more blurred the farther they get from the main object you are focusing on. The same is true of objects closer by and in the foreground—say, 3, 2, or 1 meters away from the main object. Now, if the f number is large (indicating a small diameter), you will find that objects at both 10 m and 3 m are in acceptable focus—both are not too blurred. Objects farther than 10 m or closer than 3 m, however, are blurred on the photograph. Therefore, we say that the *field* extends from 3 to 10 m and that the *depth of field* is 7 m.

Large f numbers give large depths of field, and, conversely, small f numbers give small depths of field. The reason for this is easy to understand. Consider an object whose real image is not on the film but behind it, as the red man with the umbrella in Figure 5.4. This object is *out of focus*. The green man is instead focused in front of the film, and his image is also blurry on the plane of the film. If we focus on the green man, the red man will be completely out of focus, and vice versa. The two resulting negatives will appear as shown here, with one man in focus or the other but not both in the same photo. The images are upside down, as always in cameras, and the colors are inverted on the negatives shown here. Also, the faraway green man appears smaller in the negative (magenta man).

As you can see, when taking a photograph with the aperture wide open (Figure 5.4), both objects that are nearby and further away are in focus on their planes of perfect focus. Notice how the green rays from the far object

Depth of field

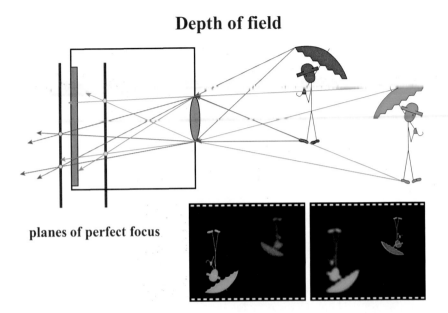

planes of perfect focus

FIGURE 5.4 ▶ The red man and the green man are close and far, respectively. Only
one of them at a time can be in focus on the film, as shown in the two negatives. Notice
that there is no aperture, or the aperture is wide open in this camera. Also note that the
images on the negatives are upside down, as they always are in real cameras, and appear
in inverted colors: cyan for the red man and magenta for the green man. These are inverse,
complementary colors. The nearby red (cyan) man appears larger than the far away green
(magenta) man on the negatives.

(green man) cross on the first plane of perfect focus. By the time they reach
the film, however, they are diverging again, and the image of the green man
will come out as blurred on the film. The red rays from the close object (red
man) also converge on their plane of perfect focus. However, since this plane
is after the film, the image the film captures will also be bigger and blurred.
When closing the aperture (Figure 5.5), *the planes of perfect focus do not
move*—they are still in front and behind the film. The difference is that now
rays can only enter the lens at small angles from the optic axis. Traveling at
such acute angles, both the red and green rays diverge very little. The image of
far and close objects is small enough on the film that the focus is acceptable!
The negative in this case shows both men in focus. In Figure 5.6 we show a
striking example of the effect of a large or small depth of field on the same
field of flowers.

5.5.1 Why the *f* Number?

There is a perfectly reasonable question you may ask: "I understand why
you need an aperture to control the amount of light entering the camera, and

Depth of field
smaller lens aperture, larger depth of field

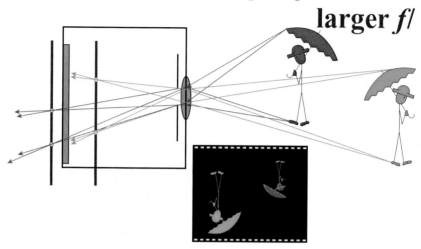

FIGURE 5.5 ► Both men are in focus on the negative when using a small aperture diameter, corresponding to a large f number.

I also understand that steps of 1.4 in diameter correspond to steps of a factor 2 in the area of the aperture, *but* why not simply label the aperture stop with the different aperture diameters instead of talking about the mysterious f number?"

The answer is that *the amount of light on the film that matters is not the total but the amount of light per square millimeter on the film.* You may think of this as the density of light rays. The f number is useful to keep this density constant. Therefore, the f number was not introduced to make the art of photography more difficult but to simplify it!

Consider the example in Figure 5.7 to clarify this point. Imagine a piece of paper with some orchids printed on it that is photographed by two lenses with focal lengths of 35 mm and 70 mm (or simply by a single zooming lens, allowing you to adjust the focal length from 35 to 70 mm—this is a very common zoom).

All lenses with the same f/ transmit the same amount of light per unit area of the film. This is quite convenient when you are using a zoom and changing the focal length of the lens because you do not need to recalculate or remeasure the exposure time with an exposure meter!

5.6 The Film

You can buy film of many types, made by many manufacturers. Let us start with black-and-white (B&W) film for simplicity's sake. Color film will be

large depth of field
(large $f/$)

small depth of field
(small $f/$)

FIGURE 5.6 ▶ These two photos were taken with different f numbers (and correspondingly different exposure times). The top image shows flowers that are well in focus in both the foreground and the background (nearby and faraway flowers, respectively). In the bottom image, a smaller f number selected only flowers at a very specific distance to be perfectly in focus. Flowers closer and farther than that distance appear blurred. Reprinted with permission from Michael Wagner.

explained in more detail in Chapter 8. All B&W films are coated with an emulsion of silver halide grains. Those grains that were illuminated turn into metallic silver when the film is developed in the darkroom. The more light strikes them, the more of them turn into silver. That is why the film produces negative images. These negatives are then inverted again to produce positive images when printed. Films differ by several properties:

- *Spectral Sensitivity*   Some films are sensitive to blue and green light only (called orthochromatic film), while others are sensitive to red light as well (called panchromatic). The advantage of ortho film is that it can be developed in the darkroom under a red "safety light" so you see the progress of the development. There are even special types of film that are made to be sensitive to the infrared or the ultraviolet wavelengths. Most black-and-white films that you can buy are panchromatic. Color films are obviously sensitive to all colors.

- *Contrast*   Suppose that you take a picture of a person wearing a relatively dark shirt that causes the light coming from the person's face to be ten times stronger than the light coming from the shirt. On the developed film will be a dark face (inverted!). If the image of the face on the

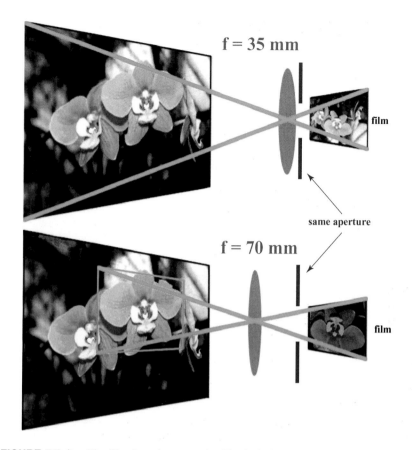

f = 35 mm

film

same aperture

f = 70 mm

film

FIGURE 5.7 ▶ The illuminated area on the film is 4 times larger for the photograph taken with the 70 mm lens. If I want the same illumination on the film as for the 35 mm lens, I must have 4 times more light coming in through the lens. Therefore, the diameter of the iris aperture behind the lens must be twice as large for the 70 mm lens as it is for the 35 mm lens. In other words, the f number must be the same! You may switch lenses or zoom in, and if you keep the same f number, you can keep the same exposure time.

film is ten times darker than that of the shirt, we say that the contrast is normal. If the face is more than ten times darker than the shirt, however, we say that the negative has high contrast, and if it is less than ten times darker, we say that the negative has low contrast.

- **Grain** The size of the original silver halide crystals and the method of development determine the final size of the silver specks that form the negative image on the film. This size is called the grain size and corresponds to the resolution or number of megapixels in digital photographs.

- *Speed* Different types of film require different amounts of light to produce the same amount of darkening on the negative. A "fast" film requires a small amount of light to darken the negative, while a "slow" film requires a lot of light.

The speed of films is measured in ISO (International Standard Organization). Currently, ISO has been adopted in all countries and has replaced both A.S.A. units (American Standards Association) and DIN units (Deutsche Industrie Norm). Figure 5.8 shows several ISO speeds that are commercially available.

Notice that ISO rating is **inversely proportional** to the amount of light required to produce a certain standard black on the negative image. Doubling the ISO number by using a faster film allows you to reduce the amount of light you need to let into the camera by 1/2.

It may seem as if it is always a good idea to buy the fastest film so you can take photographs in dim conditions of light, but you have to be careful: The fastest films (e.g. ISO 1000) have a much larger grain size. Therefore, while the photograph printed on 4" × 6" paper might appear acceptable, if you tried to enlarge it, it would appear grainy and have a low resolution. Fast films also have lower contrast (lack of deep black) and lower tonal range—that is, fewer levels of gray between black and white and fewer color tones in color photography.

Interestingly, the film speed is a useful concept not only when using film, but also for digital cameras. Most digital cameras, in fact, allow the photographer to select the ISO. Selecting higher ISO on a digital camera, just as in film photography, enables photographs to be taken in dim conditions of

Film speeds (ISO)

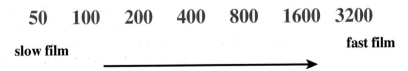

they *also* vary by factors of 2

| 50 | 100 | 200 | 400 | 800 | 1600 | 3200 |

slow film   fast film

Film speed, Sensitivity (equivalent to "more light" on film)

FIGURE 5.8 ▶ The film speed varies by factors of 2. A fast film has a large ISO number and needs less light, but has larger grain.

light. In this case the resolution is inferior, as for film, because of "binning"—that is, the amount of light/unit area on the charge-coupled device (CCD) is artificially increased by increasing the "binned" area. If 2×2 pixels are binned together, the amount of light will be four times greater and so on. In other digital cameras the resolution is not decreased, but the gain (amplification) is increased, so there is greater pixel noise.

5.7 Digital Photography

Digital photography has been embraced by artists, professional photographers and photojournalists, and, of course, by the amateur "point-and-shoot" photographers. Digital photography is no longer in its infancy, and its resolution satisfies the needs of many professional photographers and artists. Large-format digital sensors with 30 or more megapixels are now available. Film-based photography, however, and in particular large-format cameras with film sizes between $4" \times 5"$ and $8" \times 10"$ are still favored by landscape photographers. These cameras can reach an image quality considerably beyond that of the best digital cameras. The wonderful large silver-prints of nature photographs of the American South-west by Ansel Adams, printed from negatives and taken with a large-format camera, are unsurpassed.

Conventional films still have a considerable advantage when comparing the overall resolution, contrast, and tonal range of an image.

Despite the current inferiority compared to film photography, in many instances the loss in image quality is so small and the digital cameras are so convenient to use that more and more people are abandoning film photography—the authors of this book included! The conveniences of digital photography are very appealing. You don't have to buy film, develop negatives, print the photographs, store them, and in the end rarely have them handy when they are needed. It is so much easier to simply transfer the files to a computer, process, enhance and modify them as you please, and have them available at all times, stored in the small volume of a computer. We therefore look forward to further progress in the resolution, contrast, and tonal range of digital photography.

The film photography industry is certainly trying very hard to compete with the digital world. Films today are substantially better than the ones available only ten years ago. Several companies are currently experimenting with a new type of color film that could, in the near future, become ten times faster than current films without too much loss in color fidelity, resolution, or contrast. Continued research will lead to further improvements, and these may keep film photography competitive with digital alternatives. Film and digital photography now complement each other, and they may continue to do so, allowing them to coexist and provide photographers, artists, and other camera buffs with many choices for taking a picture.

Obviously, digital cameras do not require a film and instead use a charge-coupled device (CCD) to capture the image and then transfer it to a magnetic memory card. The sensitivity of CCDs, however, is still measured in ISO and can be changed by the user, so all we say about photography here applies to both film and digital photography.

5.8   Putting it All Together: Taking a Photograph

As we have seen, there are three numbers to take into consideration to capture a good photograph: the exposure time the f number, and the film speed. Figure 5.9 summarizes the values possible for each of these.

For each one of these three parameters, the most relevant variable is the amount of light on the film, or illumination I, that results. The arrows in Figure 5.9 indicate the directions in which I increases as you change each variable. Again, whenever varying the *exposure time*, the *f number,* or the *film speed* by *one step, I* varies by a factor of *two*. This is quite convenient because it greatly simplifies all the calculations to be done before taking a photograph.

Once you have the film in the camera, the film speed cannot be changed. You are left with only the exposure time T and the f number $f/$ to alter, or

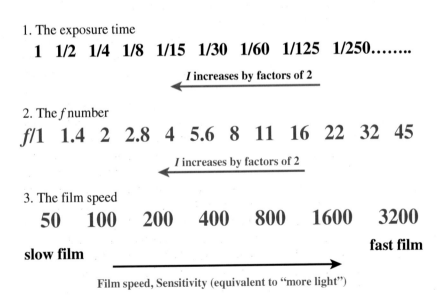

1. The exposure time

   1   1/2   1/4   1/8   1/15   1/30   1/60   1/125   1/250........

   ← *I* **increases by factors of 2**

2. The f number

   $f/1$ **1.4   2   2.8   4   5.6   8   11   16   22   32   45**

   ← *I* **increases by factors of 2**

3. The film speed

   50   100   200   400   800   1600   3200

   slow film ————————————————————→   **fast film**

   Film speed, Sensitivity (equivalent to "more light")

FIGURE 5.9 ▶ Taking a photograph, while considering all relevant parameters: the exposure time, the f number and the film speed.

to "play with" if you really like photography! The total amount of light I that falls on the film depends on

$$I = \frac{L \times T}{(f/)^2}$$

where L is the amount of light reflected by the object being photographed and transmitted by the camera lens.

Remember, the f number of the lens is given by $f/ = \frac{f}{Diameter}$.

For an example, let us assume that we are taking pictures outside. Since we cannot change the amount of sunlight reflected from the object toward the lens, all we can play with is T and $f/$. A picture taken at 1/60 sec and $f/8$ will have the same exposure as one taken at 1/250 and $f/4$. Here is why:

$$1/60 \rightarrow 1/125 \rightarrow 1/250 \text{ is two steps } \textit{less} \text{ light on film}$$

$$8 \rightarrow 5.6 \rightarrow 4 \text{ is two steps } \textit{more} \text{ light on film}$$

Decreasing the exposure time from 1/60 to 1/250 *decreases* the amount of light by a factor of 4. To compensate, we open the aperture by two stops, *increasing* the lens area by a factor of 4.

Are these two pictures the same? The answer is no. The picture with $f/8$ (the larger $f/$) has a larger depth of field, and a tree 3 m away and a mountain 1000 m away would both be in focus. However, a child jumping off the tree would be blurred because of the long exposure time. In the second picture with the smaller $f/$ and shorter exposure time, the child would not be blurred, but the mountain in the background would be.

When taking a photograph, a larger f number requires a longer exposure time. This is because a large $f/$ gives less light on the film. This has the advantage of an increased depth of field, but with the longer exposure time it is not possible to photograph movement effectively. A short exposure time gives less light on the film but has the advantage of capturing movement much better. Figure 5.10 summarizes all parameters. ***This diagram is fundamental for taking good photos and to solving any photography problem.***

A handheld exposure meter or the built-in exposure meter of a digital camera tells a photographer which settings will be best in the particular conditions of light measured. The exposure meter gives the photographer two numbers (or sets of two numbers): an $f/$ and a corresponding exposure time. However, the photographer is not bound to use those settings only. She may use any other combination of $f/$ and T that produces the same, ideal amount of light on the film. If she increases the $f/$, she'll obtain a better depth of field, but she will also need to increase the exposure time. The same applies to capturing movement: If she shortens the exposure time to better catch motion, she must also lower the $f/$.

$$f/1 \quad 1.4 \quad 2 \quad 2.8 \quad 4 \quad 5.6 \quad 8 \quad 11 \quad 16 \quad 22 \quad 32 \quad 45$$

Less light on film →

I increases by factors of 2 ←

BUT: increasing depth of field

$$T = 1 \quad 1/2 \quad 1/4 \quad 1/8 \quad 1/15 \quad 1/30 \quad 1/60 \quad 1/125 \quad 1/250...$$

Less light on film →

I increases by factors of 2 ←

BUT: captures movement better

FIGURE 5.10 ▶   The fundamental diagram to take good photographs.

This is why commercial handheld exposure meters always give many combinations of $f/$ and T. Depending on the meter one can set $1, 2$, or 3 steps of overexposure or underexposure. Autoexposure digital cameras (in auto mode) give only one particular combination of $f/$ and T. More sophisticated digital cameras give different combinations depending on the "scene" setting.

Furthermore, in modern cameras, one can select "exposure time priority" or "aperture priority." This means that one chooses T, and the camera automatically adjusts $f/$, or vice versa.

This combination (or combinations) of $f/$ and T represent a "good setting" for a photograph, as measured by the exposure meter. Starting from the measured "good setting," you can calculate other settings—for example, to give priority to short exposures or to large $f/$, depending on the photograph you wish to take. Starting from the numbers suggested by the exposure meter, to adjust these settings we must move ***to the right*** on the sequence of $f/$ and ***to the left*** on the sequence of exposure times (Figure 5.10).

Let us look at an example.

An exposure meter tells us that the "good settings" for taking a photograph are $f/22$ and exposure time $T = 1/30$ second.

$$f/1 \quad 1.4 \quad 2 \quad 2.8 \quad 4 \quad 5.6 \quad 8 \quad 11 \quad 16 \quad \boxed{22} \quad 32 \quad 45$$

$$T = 1 \quad 1/2 \quad 1/4 \quad 1/8 \quad 1/15 \quad \boxed{1/30} \quad 1/60 \quad 1/125 \quad 1/250...$$

However, since we do not have a tripod with us, we actually need to choose a much shorter exposure time T to avoid blurring of the photograph—say, 1/250 sec instead. (Remember, natural shaking of one's hand causes a photograph to be blurry if T is longer than 1/60 sec.)

With this alteration, we have moved three steps to the right on the exposure time sequence (three steps less light on the film, or $2 \times 2 \times 2 = 8$ times less

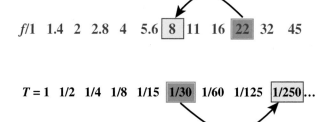

light). Therefore, we must compensate by moving three steps to the left on the $f/$ sequence (three steps more light). The new photograph is now taken with $f/8$ and $T = 1/250$ sec. The disadvantage from this change is a smaller depth of field, but an advantage is that we don't need a tripod and we can photograph action better.

Let's now consider a slightly more complicated case, in which all three parameters are changed.

The ambient light conditions determine that a good photograph is taken with 1600 ISO film, $f/5.6$, and $T = 1/125$ sec.

ISO 1600

$f/1$ 1.4 2 2.8 4 5.6 8 11 16 22 32 45

$T = 1$ 1/2 1/4 1/8 1/15 1/30 1/60 1/125 1/250...

However, we really want a better depth of field than that given by $f/5.6$. For the depth of field we want, we need to increase the $f/$ to $f/11$. However, we also ran out of film, and the only film in our bag is 200 ISO. This is a much slower film, and it requires much more light than the 1600 ISO film that we were using before. This decreased sensitivity of the film is conceptually equivalent to "less light on the film," and therefore we have to compensate for this loss by increasing the amount of light from one or both of the other two parameters.

In total, from these changes we took 3 steps less light by changing film speed and 2 steps less light by changing the $f/$. Therefore, we must compensate by increasing the exposure time T by 5 steps (5 steps more light).

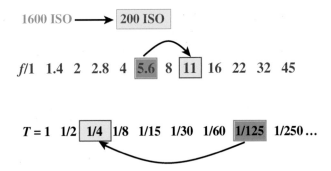

The new exposure time is T = 1/4 sec.
The disadvantage from this T is blurring when the camera is handheld.

5.9 Questions

1. What does the term *film speed* (*fast* or *slow* film) refer to? Why don't photographers always use fast film?

2. What is the meaning of the term *exposure time*?

3. What is the meaning of *f number*?

4. What is meant by *depth of field*?

5. A photograph taken with 1/60 second and $f/8$ gives the proper exposure. What f number is needed if the exposure time is changed to 1/250 second?

6. A photograph taken with 1/250 second $f/1.4$ gives the proper exposure. What f number is needed if the exposure time is changed to 1/30 second?

7. A photograph taken with 1/60 second and $f/11$ on ISO 100 film gives the proper exposure. What f number is needed if the film is changed to ISO 400 and the shutter speed is kept at 1/60 second?

8. A photograph taken with 1/60 second, $f/11$ on ISO 200 film gives the proper exposure. What f number is needed if the film is changed to ISO 400 and the exposure time is kept at 1/60 second?

9. A photographer takes three pictures of traffic on University Avenue, with the following exposure:

    (a) 1/60 sec $f/16$
    (b) 1/500 sec $f/5.6$
    (c) 1/125 sec $f/16$

    How do these pictures compare with each other?

10. You want to take a picture of your friend, and you want the landscape behind him to be sharp. Which of the two exposures would you choose, a or b?

    (a) $f/22$ 1/3 sec
    (b) $f/5.6$ 1/250 sec

11. A photograph is taken at 1/30 second and $f/2$ on ISO 100 film. What shutter speed should be used with ISO 800 film at $f/4$?

12. Most 35 mm slide projectors have lenses with a focal length $f = 5$ cm. The $f/$ of a good projector is about $f/2$.

    (a) If the screen is 3 m from the lens, how far is the slide from the lens? How large is the picture on the screen if the slide is the usual 35 mm × 24 mm?

    (b) How large is the diameter of the lens aperture? Is an $f/2$ projector lens always better than an $f/4$ projector lens? Explain.

13. A photograph taken with 1/60 s and $f/16$ gives the proper exposure. What $f/$ is needed if the exposure time is changed to 1/125 second?

14. A photograph taken with 1/60 s and $f/11$ on ISO 100 film gives the proper exposure time. What $f/$ is needed if the film is changed to ISO 400 and the exposure time is decreased to 1/125 s?

15. A photograph taken with 1/60 s and $f/11$ on ISO 200 film gives the proper exposure. What $f/$ is needed if the film is changed to ISO 400 and the shutter speed is changed to 1/30 second?

Chapter 6

Color and Color Vision

6.1 | Color

White light is a mixture of light of many different wavelengths. If you use a prism or a diffraction grating to break white light from the sun into its components, you see a sequence of colors that varies from red to violet.

The prism separates the different colors because *the index of refraction* **n** *is slightly different for each wavelength*—that is, for each color. This phenomenon is called *dispersion*. When white light illuminates a prism, the colors of the spectrum are separated and refracted at both the first and the second prism surfaces that light encounters. Refracted rays are deflected toward the normal on the first refraction and away from the normal on the second. If the prism is made of crown glass, the index of refraction for violet rays is $(n_{400\,nm}) = 1.53$, and for red rays, it is $n_{700\,nm} = 1.51$. From Snell's Law, we know that the greater the index of refraction, the more the rays are deflected. Therefore, violet rays are deflected more than red rays, as shown in Figure 6.1.

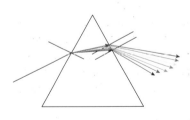

FIGURE 6.1 ▶ Dispersion from a prism.

The infinite array of colors in the real spectrum (top panel of Figure 6.2) are called *spectral colors*. Of this infinity of continuously varying colors, the human eye can distinguish fewer than one million colors. The second panel is a simplified version of the spectrum, with abrupt and **completely artificial** separations between colors. However, in speech, we do identify quite a broad range of wavelengths

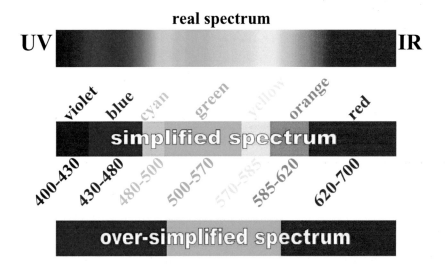

FIGURE 6.2 ▶ The colors of the spectrum. In the simplified spectrum the range of wavelengths corresponding to each color is expressed in nanometers (nm). Notice the presence of cyan, or blue-green, or turquoise, which is usually very evident in the spectrum, and the absence of indigo, a color that this author has honestly never observed in the natural spectrum.

as red, another as orange, and so on. Each range produces a specific sensation detected by our eyes, and the accurate wavelength ranges for each color are reported in the second panel of Figure 6.2. An even more extreme simplification is shown in the third panel of Figure 6.2. It is comprised of only three colors—Red, Green, and Blue—and is commonly used by computer monitors and TV sets. In these cases the three colors are combined to produce a wide range of colors, as explained in Chapter 7. You may notice that many colors you know from everyday life are not present, even in the most complete sequence of spectral colors of the top panel. Purple, magenta, and mauve are one class of colors that is missing from the spectrum. Other missing colors are all the browns and earth colors, olive and hunter greens, and all the pastel colors, such as pink, cyclamen, eggshell, aqua, sky-blue, beige, and so forth. These are mixed, low-brightness and low-saturation colors, respectively, as will be described later.

It is interesting to look at the light produced by different kinds of light sources through a diffraction grating or with a spectrophotometer. This allows us to observe what wavelengths are present and in what proportions. The light from a standard lightbulb shows all the spectral colors, but the blue and violet are less intense than in sunlight or the light from a halogen lamp (Figure 6.3). If one decreases the current flowing through the lightbulb, the filament becomes colder and the light emitted becomes dimmer and appears more yellow. When this light is analyzed with the spectrophotometer, it is evident that violet and blue have disappeared from the spectrum, explaining why the light appears to be more yellow.

On the other hand, if you use a diffraction grating or a spectrophotometer to analyze the light coming from a neon discharge tube, you will find that only certain colors are present. You see many red lines, two or three orange lines, a yellow line, two faint green lines, and a few faint blue lines.

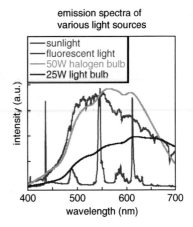

emission spectra of various light sources

FIGURE 6.3 ▶ Emission spectra from sunlight, a fluorescent ceiling light, an incandescent 25W lightbulb, and a 50W halogen bulb. Reprinted with permission from P.U.P.A. Gilbert and W. Haeberli, American Journal of Physics, **75**, 313–319 (2007). Copyright © 2007, American Association of Physics Teachers.

6.2 Color Sensitivity of the Eye

The human eye and its components were discussed at length in Chapter 4. The light-sensitive retina, found at the back of the eyeball, contains cone and rod cells. The *rods* provide low-resolution, peripheral vision and function well even in dim light, while the *cones* provide color, high-resolution, central vision and can only function in bright light. You can do an interesting experiment: Make the lighting in a room very dim, and wait until your eyes get used to the dark (dark adaptation). Now you can see things again, but you will see no colors!

We will focus on the color-sensitive cones because their functioning explains the perception of colors and color mixing. Studies on color vision indicate that there are three types of cones: Let us call them Types I, II, and III. They are sometimes called blue-, green-, and red-sensitive cones, but this is not accurate. Blue, green, and red are not the only colors to which these cones are sensitive. The sensitivity curves of the three types of cones are shown in Figure 6.4.

The *sensitivity curve* captures how the neural response changes in intensity in response to changes in the wavelength of incoming light (when the intensity of the light is kept constant). A different neural response means that a different number of nerve pulses per second is transmitted by the cones in the retina to the neuronal axons in the optic nerve and the visual cortex at the back of the brain. In Figure 6.4, the three types of cones are labeled with Roman numerals and not with usual color names in order to stress an important point: *All three types of cones are sensitive to a broad band of wavelengths.* In other words, they are sensitive to many colors. Cones of Type I are sensitive to violet, blue, and green light, Type II are sensitive to blue, green, yellow, and orange light, and Type III to green, yellow, orange, and red light.

The information that cones send to the optic nerve is simply a succession of electric pulses. The *rate* at which pulses are sent depends *both on the intensity of the light and its wavelength.*

Consider four different colors, as reported in Figure 6.5: A (cyan, with a wavelength of 500 nm), B (green, 522 nm), C (orange, 590 nm), and D (yellow, 580 nm). From the

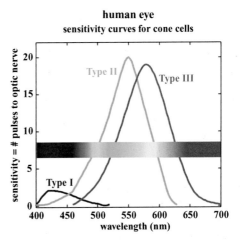

FIGURE 6.4 ▶ The sensitivity curves for the three types of cones in the human eye.

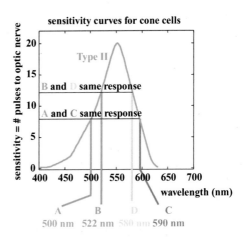

FIGURE 6.5 ▶ Response of the Type II cones to 4 different colors: A (cyan, with a wavelength of 500 nm), B (green, 522 nm), C (orange, 590 nm), and D (yellow, 580 nm).

sensitivity curves of Figure 6.4, you can read the responses of Type II cones. Notice that the responses are identical for A and C and for B and D. *If you only had Type II cones, you would not be able to distinguish cyan from orange or green from yellow.* By having the *three types of cone cells*, the brain combines the responses from each type and is able to interpret *many more colors*.

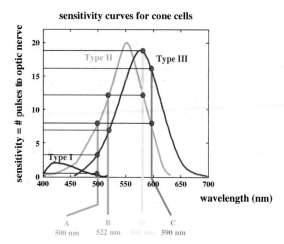

FIGURE 6.6 ▶ The same 4 colors of Figure 6.5 are now seen with all three types of cones, and can be distinguished.

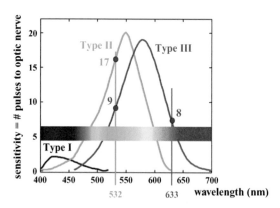

FIGURE 6.7 ▶ Take a green and a red laser pointer, with wavelengths respectively 532 nm and 633 nm. Point both of them to the same spot on a white wall. The spot will stimulate a response in your cones of 0, 17, and 17 in the three types of cones, respectively.

The magenta dots in Figure 6.6 indicate the response of all three types of cone cells. Both colors A and C generate a response of 8 on Type II cones, but at the same time, A generates only 3 on Type III cones, while C generates 17. In addition, A also stimulates a small reaction on behalf of Type I cones. No such reaction is present for C. The brain receives a set of these three numbers: the responses of cones of Types I, II, and III. Processing this information, the brain can therefore distinguish A and C. This interpretation takes place instantaneously, as you can see and distinguish the colors as soon as you see them!

One third of the brain volume is used to interpret visual input. This tells you how complicated it must be! Despite the complexity of the system, there is a problem associated with having only three types of cones: Some ambiguities occur.

Figure 6.7 shows a good example of the ambiguity generated by having *only* three types of cones. Consider two lights of wavelengths 532 nm and 633 nm. These are green and red lights. If you project these two lights onto a white screen (you *add* them), the brain will interpret them as the *sum* of the responses of all three types of cones. The response of Type I is negligible. For

Type II it is 17, and for Type III it is $9 + 8 = 17$ (see the magenta dots in Figure 6.7).

Let us now look at a third light of wavelength 570 nm, which is yellow (Figure 6.8). This generates responses of 17 for both Types II and III. These responses are identical to the ones obtained previously from the sum of green and red. That is why when you add red and green lights, the eye perceives the resulting light as yellow even if no yellow wavelengths are present!

Analyzed with a spectrograph, the mixed yellow has two wavelengths (532 nm and 633 nm), and the spectral yellow has only one wavelength (570 nm). Despite this spectroscopic difference, *the eye cannot tell the difference between one light of 570 nm and the sum of two lights of 532 nm and 633 nm!* Perhaps if we had 4 or more types of cones we would not have such ambiguities in our vision. On the other hand, more types of cones would require more volume of the brain to be used for vision, and this might have a negative effect on our other skills. Even worse, if we did not have these ambiguities, we would not be able to mix colors as easily! Then we could not have

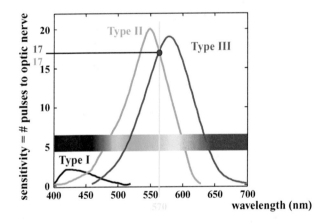

FIGURE 6.8 ▶ Yellow light at 570 nm generates a response of 0, 17 and 17 in the three types of cones. This is identical to the sum of the red and green lights in Figure 6.7. This is why red plus green gives yellow. Note that the red and green to be added to give yellow do not have to be single-wavelength lights like the laser pointers of this example. Broad-spectrum red or green beams, as those filtered through colored glass windows in Gothic cathedrals, or simply flashlights beaming through red and green filters to the same spot produce yellow. If you are using sun as a source of light for this experiment all beams are parallel and you can't make the red and green lights overlap. You may use two converging lenses, or one small mirror to direct both red and green beams to the same spot.

such a simple coding system for RGB color TV, computers, and projectors, and we would have much more difficulty mixing pigment colors.

Cats have very poor color vision because they have very few cone cells. Their eyes are optimized in several ways for night vision. Since cone cells only function in bright light and cats are nocturnal animals (more active at night), they have evolved to have a majority of rod cells that work well in dim illumination. Other differences between our eye and a cat's eye are summarized in Figure 6.9.

Cat's eyes have *vertical slits* as pupils that can open up much more (at night) and more easily than the radial and circular iris muscles of a human's circular pupil. A larger aperture has higher luminosity because more light can go through the lens, but it also has the disadvantage of a small depth of field. The lowest $f/$ (fastest lens, highest luminosity) for camera lenses is $f/1$, for the human eye $f/2.1$, for an owl, another night predator, $f/1.3$, and for a cat, the f number is $f/1$. The cat also has a *tapetum lucidum* (Latin for "shiny carpet"), a layer of reflecting cells at the back of the retina. The retina is semitransparent, and it does not absorb all the light that illuminates it. Part of the light is transmitted through the retina and is simply lost. Night animals have the tapetum to reflect that portion of the light back into the retina. The tapetum is located immediately behind the retina, and therefore the reflected rays do not go out of focus much, allowing the rods in the retina to have a second chance to detect them.

Many people believe that snakes can see infrared light, but this is not true. Snakes do sense the heat irradiated (infrared) by their prey at night through

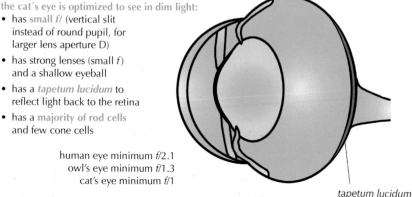

the cat's eye is optimized to see in dim light:
- has small $f/$ (vertical slit instead of round pupil, for larger lens aperture D)
- has strong lenses (small f) and a shallow eyeball
- has a *tapetum lucidum* to reflect light back to the retina
- has a majority of rod cells and few cone cells

human eye minimum $f/2.1$
owl's eye minimum $f/1.3$
cat's eye minimum $f/1$

tapetum lucidum

FIGURE 6.9 ▶ A cat's eye.

FIGURE 6.10 ▶   Bees do not see red, but see in the ultraviolet.

their tongue, but they do not *see* it. One exception is the rattlesnake, which has an infrared pinhole system that allows it to actually see infrared light. Dogs and flies see colors just as humans. Bees do not see red but do see ultraviolet light, which humans cannot see. To attract bees, some flowers have structures on their petals that are only visible in the ultraviolet, and we cannot see (Figure 6.10).

Many other insects perceive UV light, as do birds, lizards, turtles, and many fish. All these animals have a fourth type of cone in their retina, which we humans completely lack. We do not have, by any means, the most sophisticated eye in the animal kingdom!

A more direct way of studying how our eyes perceive color is to analyze the addition of lights, or ***additive color mixing***. This is described in detail in Chapter 7.

Before starting to add lights, let's introduce a few criteria and definitions that will be useful to understand how colors are mixed.

6.3 Physical and Psychological Color

The physical, measurable color and the perceived, psychological color are different. An example of a ***physical color*** is spectral yellow in sunlight or from a sodium lamp. This is a color which is comprised of only one wavelength, also known as a ***spectral color***.

A *psychological color*, or *hue*, can be *a single wavelength or a superposition of different wavelengths*. Think of yellow: The color we see can be a spectral yellow with a dominant single wavelength of 570 nm or the superposition of red and green lights with two different wavelengths, it could even be a broadband spectrum that appears to have the same color as 570 nm spectral yellow. All these yellows are *metamers*: they have different physical origins but generate the same perceived color for the normal eye.

Before we can discuss color further, therefore, we need to establish an accurate definition of *color*.

6.4   Color: Hue, Saturation, and Brightness

The physical parameters of light are not sufficient to accurately describe the color we see. A psychological response to the physical stimulus generates the visual sensation. Color is therefore best described by *psychophysical parameters*, which we will call *hue, saturation*, and *brightness*. Here all three, and their synonyms and antonyms, are defined.

Given any color, a matching color that appears to the normal retina identical to the color in question is obtained by combining light of a single wavelength with white light.[1] The wavelength of the matching color is called *dominant wavelength*. The corresponding attribute of visual sensation is called the *hue* (or simply *color*). One minus the fraction of white light is called *saturation*. Saturation is also known as *purity*. High-saturation colors require a small fraction of white in the matching color, low-saturation colors require a large fraction of white. The latter are pale pink, sky blue, pale yellow, beige, and all those colors commonly called *pastels*. The adjectives saturated and *pale* are therefore antonyms. To match those colors, the matching color requires a large fraction of white. Some authors define saturation in paints and pigments to be the parameter related to the amount of black, but we do not see a reason for such distinction. *Brightness* (also known as *lightness*) is the parameter of a color according to which an area appears to emit more or less light. The adjectives *bright* and *dark* (or *light* and *dark*) are therefore antonyms. In the literature there is a preference to use the term *brightness* for luminant (light-emitting) sources and *lightness* for illuminated (light-reflecting) surfaces.

Hue, saturation, and brightness are considered psychological parameters by other authors and are deemed to be nonmeasurable. The computer industry, however, has made the decision of adopting these three terms (abbreviated

---

[1]This definition does not apply to magentas and purples. Because magentas and purples are non-spectral colors, they can be matched by the wavelength of their complementary color. See non-spectral colors in Chapter 7.

HSB) to quantitatively and reproducibly describe colors. We therefore think it wise to adopt these three terms that, although not perfectly in line with previous physics publications, are in good agreement with every day language and with current technology.

Psychologically, pure yellow (0% white) appears to have *low saturation* and seems to be lighter, while red, green, and blue appear to have *high saturation,* and seem to be fuller colors compared to the paler yellow. This is due to the higher sensitivity of the Types II and III cones in the yellow region (either single wavelength or broad-band yellow at 500–700 nm), compared to other wavelengths. There is no physical correspondence to this impression, which is entirely due to the physiological response to light and colors, not to the actual brightness or saturation of the physical color observed.

The *intensity* of the *physical color* corresponds to the **brightness** of the **psychophysical color.** The perceived, psychological brightness cannot be measured. It is proportional to the logarithm of the intensity, but it varies dramatically from person to person, and even for the same person, depending on the surrounding illumination and the *adaptation* of the eye to both color and light intensity. If the eyes are dark-adapted—that is if they've been in darkness for a few minutes or longer—when the light is turned on again, a dim color may appear brighter than it does when the same eyes of the same person have not been dark-adapted. The physical intensity of that color, which is a measurable quantity, remained unchanged in both conditions.

The perceived brightness of a color also depends on the surrounding illumination. A snowball indoors is perceived as white, and a piece of coal outdoors under sunlight illumination is perceived as black, even though the snowball indoors scatters much less light than the coal outside! Brightness depends on the environment surrounding the object being observed. In bright sunshine a black object can still reflect a large amount of light, but since everything else around it reflects more, it is perceived as black. This effect does not only take place under intense sunshine. Observe the checkerboard in Figure 6.11.

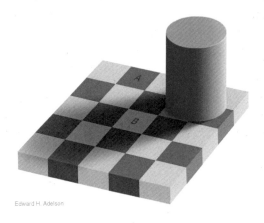

Edward H. Adelson

FIGURE 6.11 ▶ The checker shadow illusion. Reprinted from Edward H. Adelson. Copyright © 1995.

The squares marked A and B are the same shade of gray. To check this, cover the figure with a piece of paper with two holes punched to show only the areas A and B. B appears much lighter than checker A because it is surrounded by darker shades of gray. You can find this and many other similar illusions on Edward H. Adelson's website at MIT: *http://web.mit.edu/persci/gaz/.*

<table>
<tr><td>**6.5**</td><td>**Light Interaction with other Objects**</td></tr>
</table>

The light illuminating objects can be any of the following:

- Absorbed
- Specularly reflected
- Scattered or diffusely reflected (synonyms)
- Transmitted and refracted
- Combinations of the above

If light from a light-bulb illuminating an object is completely ***absorbed*** by that object, no light is reflected, such as when the object appears black. If only some wavelengths of light are absorbed and others are ***scattered***, the object will appear ***colored***. If light is ***specularly reflected***, such as when the object is an aluminum or silver mirror, the light does not change color at all and is reflected in one very specific direction, defined by the law of reflection ($i = r$, as described in Chapter 2). In this case, the virtual image of the object can only be seen from a very specific position. Conversely, when light from the light-bulb is ***scattered*** by the object, it bounces back in all directions, and the object can be seen from any angle. If the object is transparent, light will be also ***transmitted and refracted***. The most common situation is to have a ***combination*** of absorption and scattering of different wavelengths. This generates the colors of all objects surrounding us. If the objects are smooth and shiny, they are also specularly reflecting a portion of the light. Let us now look at the specific details of these situations.

<table>
<tr><td>**6.6**</td><td>**Scattering or Diffuse Reflection**</td></tr>
</table>

Scattering (or *diffuse reflection*) takes place when light illuminates an object that does not have a perfectly smooth and flat surface. We have seen that in specular reflection, such as reflection from a mirror or a shiny smooth object, the angle of incidence is equal to the angle of reflection. In diffuse reflection this is not the case. In fact, an object can have any kind of surface roughness, texture, or finish, and we can usually see it from all angles. Light illuminating any non-mirror object is *scattered in all directions* away from the object. The scattered light rays are then visible from all angles, all around the object.

In most cases, the conditions needed for specular reflection are satisfied at the microscopic level, but there are so many different microscopic surface orientations on a rough, non-mirror-like surface that the end result is light bouncing off in all directions.

For simplicity's sake, we can divide all the colors in white light into three spectral ranges: blue, green, and red, as shown in Figure 6.2. We are therefore neglecting the spectral orange, violet, yellow, and blue-green or cyan. This is an oversimplification in some cases, but it is useful and can be applied to most real-life situations and produce accurate results. In the examples following and throughout the remainder of this book, we will use this simplified spectrum.

A *black* object, such as a piece of lava rock, *absorbs most of the light intensity* when illuminated by white light. A very small amount of light is scattered, and from that small amount we can tell the shape and size of the rock. Our eye can see the small amount of light, perceive its very low brightness compared to the colors of the surrounding environment and to the illuminating light, and interpret the color as black. All colors in white light (red, green, and blue) are absorbed equally by the black rock, and a small proportion of each one, in equal amounts, is scattered.

A *white* piece of paper or a snowball behaves in a very similar way: All colors are equally reflected. The difference is that in this case, *no light is absorbed. All the illuminating light is scattered.* The blue, green, and red components are reflected in equal amounts, our eyes see the high brightness, and both the paper and the snowball appear white.

All colored objects absorb certain wavelengths and scatter others. The apple in Figure 6.12 appears red because, when illuminated by white light, it absorbs green and blue and scatters red light only.

A lemon appears yellow because it absorbs blue light and scatters red and green light. Our eyes perceive these red and green wavelengths, we add them together, and interpret the lemon's color as yellow. We can also think about the color perceived in another way: *An object appears as the color complementary to the color it absorbs.* In other words, a lemon appears yellow because it absorbs blue, the complementary color to yellow. Furthermore, the lemon appears shiny on the surface (see Figure 6.13) because a small portion of the light (white light!) is specularly reflected. The same goes for the apple in Figure 6.12.

FIGURE 6.13 ▶ A lemon appears yellow because it absorbs blue, the color complementary to yellow. It also reflects white light in the area that satisfies the condition for specular reflection.

FIGURE 6.12 ▶ An apple is red because it absorbs green and blue, and scatters (diffusely reflects) red. It is also shiny, thus in the areas accurately positioned for the apple surface to undergo specular reflection with respect to the observer's eyes, it appears as a mirror: white light illuminates it, white light is reflected.

The *complementary colors* are

cyan ⇔ red

magenta ⇔ green

yellow ⇔ blue

They appear on opposite sides of the diagram in Figure 6.14. There is a simple rule to memorize the complementary colors: The most commonly used sequences are RGB and CMY (e.g., RGB monitors or CMY ink cartridges). Colors in the same positions in these sequences are complementary.

It is interesting to note that Descartes in 1630 attributed the color of an object to a change in the illuminating light when this is reflected by the object. This is exactly what we now know is the correct interpretation of color. A red apple is red because it absorbs all colors in the illuminating light except for red and scatters red. The illuminating and scattered lights, therefore, are different, exactly as Descartes intuition predicted.

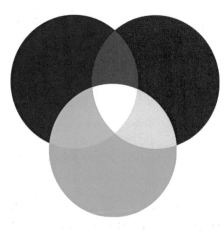

FIGURE 6.14 ▶ Three colored light beams, red, green and blue, overlap two at a time to form cyan, magenta and yellow. Colors at opposite ends of this diagram are complementary colors: red and cyan, green and magenta, blue and yellow. Where all three colors overlap the resulting color is white.

Scattering also depends on the size of the scattering illuminated particles. If scattering particles are larger than the wavelength of light, the scattering is the same for all wavelengths (all light colors). If the particles are smaller, the scattering is wavelength dependent. Let us see a few cases in which scattering produces interesting effects.

Water is almost completely transparent. But then, why does water in a Baroque fountain or a waterfall appear white? This phenomenon is an effect of scattering. The water particles—that is, the droplets in the waterfall—are much larger than the wavelength of visible light (400–700 nm). Therefore, light scattering is wavelength independent. This is also called Mie scattering. If white light is illuminating the waterfall, white light is scattered in all directions by the water droplets. Since all colors (wavelengths) composing white light are scattered equally, we see the waterfall as white. The same wavelength-independent scattering is responsible for the whiteness of the clouds. The ice crystals and water droplets that form clouds are larger than the wavelengths of light, so they appear white.

Let's talk some more about scattering: Why is the sky blue? Air, with very good approximation, is transparent, so why does a lot of air (the sky) appear blue? In this case, the air particles (76% nitrogen molecules and 21% oxygen molecules) are much smaller than the wavelength of light, and this causes the scattering to be wavelength dependent. This is also called Rayleigh scattering. The amount of light undergoing Rayleigh scattering is proportional to $1/\lambda^4$, where λ is the wavelength of light. Therefore, blue light, which has the shortest wavelength (400–500 nm), is scattered more than green (500–600 nm) and red light (600–700 nm), and the sky appears blue. Although blue is scattered more, all wavelengths are scattered. This is why the sky is light blue (in fact, sky blue = unsaturated blue = white + blue).

Sometimes the sky is very light blue; other times it is a much more saturated blue. Wavelength-independent Mie scattering from water droplets in mist or haze desaturate the sky color. In dry climates the sky is always more saturated. Figure 6.15 clarifies the geometry of direct sunlight and scattered light from the sky.

Imagine light from the sun coming toward you. Along the way it encounters air molecules in the atmosphere and bounces off of them at all angles. Then, the blue light, which was scattered more than green and red at the first bounce, continues to bounce off other air molecules again and again. The whole sky becomes illuminated homogeneously as blue. In the absence of an atmosphere to scatter light, the sky would not be illuminated at all, and it would appear black. This is the case for the moon, which does not have an atmosphere. On the moon, with no light scattering from air molecules, areas that are directly illuminated by the sun (in line of sight of the sun) are extremely bright and the shadows are completely dark.

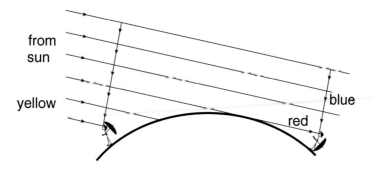

FIGURE 6.15 ▶ Formation of the blue sky and red sunrise or sunset by scattering in the atmosphere. Redrawn and adapted with permission from K. Nassau, The Physics and Chemistry of Color, Wiley, New York, 2001.

Fog and mist, as clouds, are formed by water droplets larger than the wavelength of light (usually 1000 nm or larger), so they scatter wavelength-independently and appear white. Manufacturers of automobile fog lamps, however, sometimes make these produce yellow light, so that less scattering occurs from the smaller droplets that are also present.

Why is the ocean blue? It is by no means true that the ocean is blue because it reflects light from the sky. This can easily be proved on the surface of still water (e.g., in a swimming pool) with an external source of light and a polarizer filter. It is also not true that blue in water and ice is a result of scattering. Although some scattering is certainly present, the amount of scattering-generated blue is orders of magnitude lower than the color generated by water absorption. The answer is simple: The ocean is blue because water is blue, or cyan, to be perfectly accurate. The red part of the spectrum is absorbed by water, leaving blue and green (which are not absorbed) to be scattered. Not by chance, aqua is a blue-green color (blue-green = cyan and is complementary to red). Water absorbs red, so it appears cyan. The same is true for the pale cyan color of bulk ice, such as in glaciers or icebergs. A green color in water or ice usually indicates the presence of algae, which contain a green pigment (chlorophyll).

When the background is white, for example, white sand, a white marble fountain, or a white bathtub—water appears cyan because it absorbs red. Figures 6.16 and 6.17 show these examples.

In summary, from all these examples we conclude that *the color of an object depends on many different aspects*, including the following:

- The wavelength and intensity of the illuminating light
- The wavelength of light scattered or transmitted by the object
- The color of surrounding objects
- The absorption or scattering of substances in the path of light

FIGURE 6.16 ▶ Colors of the Pacific Ocean from the Huahine Island in French Poly-
nesia. At the center of this photograph there is a white sand bank underwater. The shallow
water above it, therefore, appears an unsaturated cyan—that is, white (from the sand) and
cyan (from the water itself). Outside the reef, where the water is deeper and rougher, the
wind produces white caps, foam, and isolated air bubbles, which scatter light and keep it
from penetrating into the water. As a result, the water color is darker blue. In the foreground of this photograph the water is also deeper but not rough and bubbly, so it has an
intermediate brightness.

FIGURE 6.17 ▶ Water in the Barcaccia Fountain in Piazza di Spagna, Rome, Italy,
designed by Pietro Bernini, Father of Gian Lorenzo. Shallow water against the white marble
background appears cyan because water absorbs red.

6.7   Questions

1. Visible light ranges in wavelength from 400 to 700. In what units are these numbers expressed?

2. What wavelengths is red associated with in a simplified spectrum? Green? Blue?

3. What are the eye's color sensors called?

4. Draw graphs of the relative sensitivity of the three types of color receptors in the retina, showing how the color sensitivity of each changes with wavelength.

5. Are "green sensors" only sensitive to green light? Explain.

6. What are complementary colors? Which ones are they?

Chapter 7

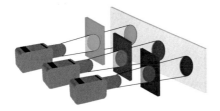

Additive Color Mixing

Additive color mixing works for ***mixing lights***. The rules of additive color mixing apply to color TV, computer monitors, projectors, and stage and disco lights. These rules *do not* apply when mixing pigments and paints. The three projectors shown here all emit white light. Each projector is illuminating one filter. The color each filter transmits onto the screen is the same color as the filter. The red filter transmits red light, the green transmits green light, and so on. We can now mix these three colored lights by projecting them onto the same location on the white screen. By overlapping them, we can obtain almost any colors. On a TV or computer monitor, instead of overlapping the three colors, they are positioned side by side in small areas called *pixels*. The pixels are small enough that from a normal viewing distance the eye perceives the additive mixture of red, green, and blue. Look at your TV or computer screen with a magnifier and you will see the individual pixels. Let us start with the additive primary colors.

7.1 | Primary Colors

The three filters in front of the three projectors of ***identical intensity*** are red (R), green (G), and blue (B). If all three colored lights are projected on a white screen, partly overlapping, they produce the additively mixed colors displayed in Figure 7.1. As shown, adding blue light to red light produces magenta (M) (B + R = M), while adding B + G = cyan (C) and adding R + G = yellow (Y). Where R, G, and B overlap, white is obtained. R, G, and B are therefore three ***additive primary colors***. The definition of primary colors is

(1) one primary cannot be matched by a mixture of the other two, and (2) all three, when combined in proper proportions, produce white.

There are many other combinations of three colors that correspond to this definition. An example can be seen in Figure 7.1: C, M, and Y. These could also be primaries since each color cannot be made by a mixture of the other two, and, when combined in proper proportions, they produce white. As we

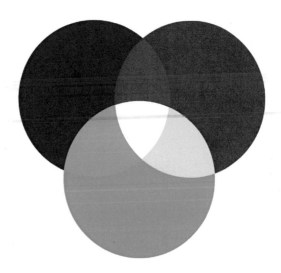

FIGURE 7.1 ▶ The most frequently adopted additive primary colors are R, G, and B.

will see in more detail later (Figure 7.9), however, when mixed, they can only generate a limited number of colors. Therefore, they are not the most useful additive primaries.

Two *complementary colors* are *two colors that, when additively mixed in appropriate proportions, produce white*. Additively mixed implies that we are adding lights of these two colors, as with the above projectors and colored filters (not paint!). Note in Figure 7.1 that the complementary colors appear on opposite sides of the diagram and that the complement of each primary color is the combination of the other two primaries.

7.2   Adding Primary Colors

It does not seem possible, at a first sight, that we can mix R, G, and B to obtain the two colors reported in Figure 7.2: chocolate brown and cyclamen pink. We will describe in detail how to do this quantitatively. For now, it will suffice to say that there are accurate combinations of R, G, and B that produce these two colors. These are: 66% R, 33% G, and no B. Then to obtain brown, we must also decrease the intensity of the light. Pink is a mixture of approximately 40% R, 30% G, and 30% B. In fact, *almost any color can be obtained with a combination of R, G, and B* lights.

Let us illuminate a white screen using a projector with a light of an arbitrary color (stimulus S). In Figure 7.2, S is brown or cyclamen pink. To do this, we can place on the screen a colored sheet of paper and illuminate it with white light or have a colored filter in front of the white light projector. Let us now illuminate an empty part of the white screen in Figure 7.2, with R, G, and B lights, all pointing to the exact same location, so they overlap. If we vary the intensities of these three lights, we can obtain a match for brown or pink. A match of the color can be represented by the equation

FIGURE 7.2 ▶ Chocolate brown and cyclamen pink can be matched exactly by a combination of R, G and B.

$$S = aR + bG + cB$$

In this equation, a, b, and c are the intensities of R, G, and B lights, respectively, that are needed to match the stimulus color. The units or range of values used for the intensities (a, b, c) are of no interest. The fractions of R, G, and B that make up the light are the only numbers that matter when defining the color of the light. We can calculate what fraction r of the light is red:

$$r = \frac{a}{a + b + c}$$

And similarly what fraction g is green:

$$g = \frac{b}{a + b + c}$$

These two numbers, r and g, define the mixture of colors projected on the left spot.

In the cyclamen pink example we have

$$S = 250\,R + 170\,G + 200\,B$$

The numbers reported here for a, b, and c (250, 170, and 200) are the intensities of R, G, and B used on the computer to generate the cyclamen pink in Figure 7.2, each of which can vary between 0 and 255. The total Intensity I is the sum of the intensities for R, G, and B:

$$I = a + b + c = 250 + 170 + 200 = 620 \text{ units.}$$

Remember, the units in the intensity are irrelevant because the only numbers that matter to define the color are the ratios, or the fractions, of R, G, and B. These ratios are unitless. On a computer with 8-bit color images, the values of a, b, and c vary between 0 and 255. There are, therefore, $256 (= 2^8)$ intensity levels for R, for G, and for B. You can try the preceding numbers on your computer and verify that you really do obtain cyclamen pink. The ratios we need to calculate are

$$\text{The fraction of red r} = \frac{250}{620} \approx 0.40$$

$$\text{The fraction of green g} = \frac{170}{620} \approx 0.28$$

$$\text{The fraction of blue b} = \frac{200}{620} \approx 0.32$$

Notice that $r + g + b = 0.4 + 0.28 + 0.32 = 1$

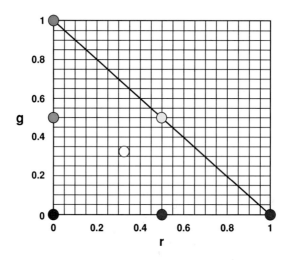

FIGURE 7.3 ▶   **The color triangle.** The fraction of red and green may vary between 0 and 1. The fraction of blue does not need to be plotted because it is always the complement to 1 of the other two. If the fraction of green is 0, the colors on the horizontal axis are B when $r = 0$, R when $r = 1$, and M when $r = 0.5$ (midpoint between R and B). On the vertical axis ($r = 0$), we will have G when $g = 1$ and C when $g = 0.5$ (midpoint between G and B). Y is between R and G, or when r and $g = 0.5$. White is at the point where both r and g are 0.33. The color triangle here shows all these points (R, G, B, C, M, Y, W) and the corresponding colors.

This is always true! *The fractions r, g, and b always add up to 1*. Therefore, we do not need to calculate all three fractions. Calculating the fraction of red and the fraction of green is sufficient because the fraction of blue is always the remainder ($b = 1 - r - g$).

We can now plot the fraction of red on the horizontal axis and the fraction of green on the vertical axis of a diagram such as the one in Figure 7.3. This diagram is called the chromaticity diagram or the *color triangle*.

7.3 The Color Triangle

We can find the coordinates of any mixed color on the color triangle. The color triangle describes hue and saturation, but is silent about brightness. The color triangle rule is the following: When *adding two colors, the resulting color is always on the line joining the two colors*.

This is why C is in the midpoint between B and G, M between B and R, and Y between R and G. The color triangle in Figure 7.4 provides a simple way of mixing additive primary colors to obtain a desired color, or to anticipate exactly what color will result when two colors are mixed in known proportions.

Any mixed color is the sum of the corresponding spectral color plus white. A high-saturation color contains no white, while a low-saturation

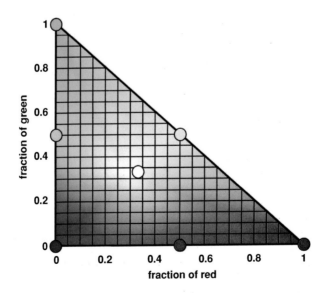

FIGURE 7.4 ▶ The color triangle here is filled with all the colors that can be obtained by mixing R, G, and B.

(pastel) color contains lots of white. In Figure 7.5 along any line starting from white and going toward the edges of the triangle, the hue of a color does not change. The only thing that changes is the saturation. Saturated colors are near the edges of the triangle, while low-saturation colors are near the center.

Once the fractions of red and green of a color are identified, all we need to do to find the exact hue of the color is graph the point (with coordinates r and g) on the color triangle, draw a line from white to that point, extend the line until it reaches the edge of the triangle, and identify the color where the line crosses the edge. We then identify the saturation of the color by measuring how far the point is from the edge.

Take the example of cyclamen pink. We found that r = 0.4 and g = 0.28. We graph the point with these coordinates on the triangle in Figure 7.6 and report it as a white circle, extend a line from white to this point, and then

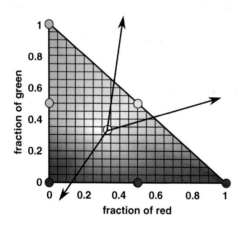

FIGURE 7.5 ▼
Along the arrow lines drawn on this color triangle only the saturation changes: it increases as the color is farther away from the center, and contains less white.

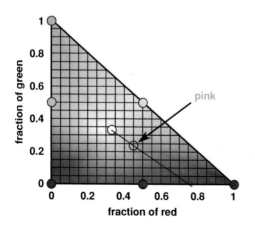

FIGURE 7.6 ▼
Pink, or low saturation red, is the color we find at r = 0.4 and g = 0.28.

continue the line until we reach the edge. The color at the edge (the corner in this case) is red. The point corresponding to our color is much closer to white than to red; we therefore conclude that the color is a low-saturation red. This is known outside of physics as pink. If the distance from white to pink is five times smaller than from white to red, the pink is in fact 1/5 red and 4/5 white.

When mixing two colors, we must first identify them on the color triangle. Then we must draw a line between them. If the two colors are mixed in equal amounts, the resulting color is at the midpoint between the two original colors. If there is more of one color than the other, the resulting color is closer to the more abundant of the two.

According to the color triangle rule, when mixing two colors, the resulting color lies along the line connecting the two. In Figure 7.7 we mix 1 part of high-saturation orange and 2 parts of high-saturation green; the resulting color lies 1/3 of the way from green and 2/3 of the way from orange along the line connecting the two colors, and the resulting color is a high-saturation lime green (thicker white circle). Similarly, if we mix 2 parts of Y and 1 part of C, we obtain low-saturation lime green. Imagine this point on the color triangle.

By now, it should be clear why R, G, and B are usually chosen as the additive primary colors: RGB provide the largest mixing potential. In other words, the largest number of colors can be obtained from mixing RGB primaries—all the colors in the color triangle. If we chose red, blue, and yellow as additive primary colors, we could only obtain the colors within the

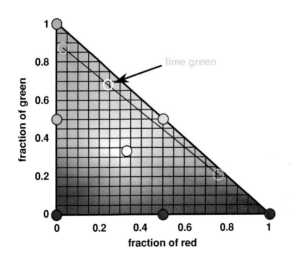

FIGURE 7.7 ▼
Mixing two parts of green and one part of orange, we obtain lime green. The resulting color is on the line connecting the two original colors, 1/3 of the way from green, 2/3 of the way from orange.

triangle enclosed by the three points where these three colors lie on the color triangle, as shown in Figure 7.8. We could obtain R, O (for orange), Y, W, B, M, and all the colors intermediate between these, but we could never obtain G or C. This is shown in Figure 7.8. A choice of C, M, and Y would be even more limiting because we could never obtain red, or blue, or green. This case is shown in Figure 7.9.

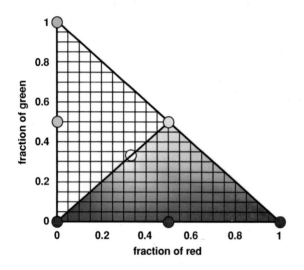

FIGURE 7.8 ▼
If we used RYB as primaries, only the colors within the RYB triangle could be obtained, to the exclusion of all greens and blue-greens.

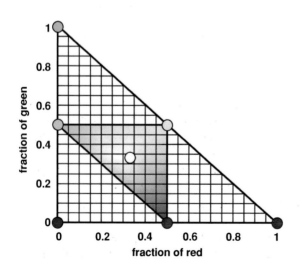

FIGURE 7.9 ▼
If we used CMY as primaries, only the colors within the CMY triangle could be obtained by mixing. No R, G or B could ever be obtained by mixing CMY in any proportions.

7.4   Low-Brightness Colors

Are *all* colors represented on the color triangle? Where are the olive greens, the grays, the browns, beige, and black? These colors are indeed present on the triangle, but since *the color triangle does not represent the color brightness*, these low-brightness colors are not immediately evident. The color triangle only shows the color's hue and saturation but not the brightness. Olive green is nothing but low-brightness green, while brown is low-brightness orange, gray is low-brightness white, and black is the lowest-brightness white. Black can actually be the lowest (or zero) brightness of any color. Black could therefore be anywhere on the color triangle. For these colors, the RGB amounts on a 0–255 level basis are as follows:

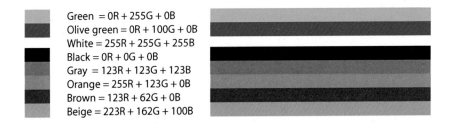

$$Green = 0R + 255G + 0B$$
$$Olive\ green = 0R + 100G + 0B$$
$$White = 255R + 255G + 255B$$
$$Black = 0R + 0G + 0B$$
$$Gray = 123R + 123G + 123B$$
$$Orange = 255R + 123G + 0B$$
$$Brown = 123R + 62G + 0B$$
$$Beige = 223R + 162G + 100B$$

The color bars next to each color description were obtained using these accurate values. Please try them on your own computer. Any graphics program (Adobe Photoshop®, Adobe Illustrator®, PowerPoint®, Canvas®, etc.) will allow you to enter the RGB numeric values you desire when using a "color picker" to select a color. These colors are also reported on the color triangle of Figure 7.10.

7.5   Spectral Colors

Not all the spectral colors can be matched by mixing RGB. *The colors obtained mixing RGB primaries are less saturated than their corresponding spectral colors* (dominant wavelengths). This is particularly evident in the *blue-green* region, also called *cyan* or *turquoise*.

Let us project on a screen in spot #1 a combination of RGB and on spot #2 spectral cyan (480 nm). Let us try to match the color of spot #2 with a combination of RGB in spot #1. As shown in Figure 7.11, in spot #1, we would project blue and green, with wavelengths 420 nm and 550 nm, respectively. We need to use our imagination here because the color in spot #2 was produced on a computer that works additively and is therefore not as saturated as the true spectral cyan with 480 nm. Please imagine that the cyan in spot #2 has a much higher saturation than the mixed cyan of spot #1.

To obtain a perfect match, we need to produce the same response in the cone cells of our eye. We will find that the only way to obtain a match is to shine some red light on spot #2 and then match the new color with blue and

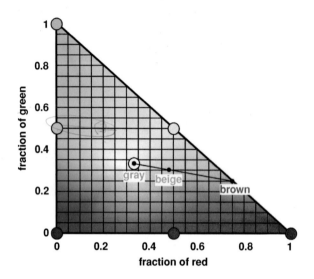

FIGURE 7.10 ▶ Olive green, brown, and gray are shown as dark dots in the color triangle. They are in the same positions as green, orange, and white, since these are low-brightness colors, and the color triangle does not reveal brightness. Beige is low-saturation brown, and it lies along the line between white and brown (or orange).

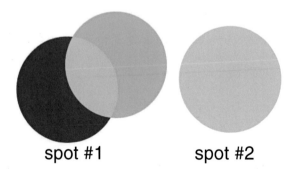

FIGURE 7.11 ▶ A combination of R and B (spot #1) trying to match spectral cyan (spot #2).

green light on spot #1. The reason for this is clearer if we look again at the sensitivity curves for the cones and their responses to the stimuli on spots #1 and #2. These are reported in Figures 7.12 and 7.13.

If we try to obtain the same result as from C by adding B and G, we must first adjust the B intensity to give the correct Type I response (about 1 unit in Figure 7.12). We then add enough 550 nm light to get the desired Type II response of 4 units. Since 550 nm *also* stimulates Type III cones a lot (about 3/4 as much as Type II; in fact at 550 nm for 19.5 units on Type II we also get 14 units on Type III), however, this means we have added too much red—namely about 3 units (3/4 of 4; Figure 7.13).

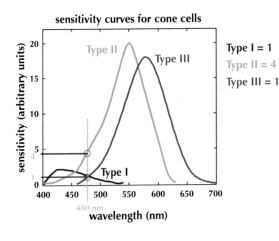

FIGURE 7.12 ▼
The responses of the three types of cones to the stimulus of spot #2: spectral cyan with 480 nm wavelength.

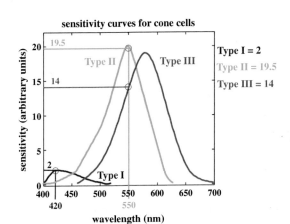

FIGURE 7.13 ▼
The responses of the three types of cones to the stimulus of spot #1: a mixture of primary G and B with 550 and 420 nm wavelengths, respectively.

The addition of R to B + G is equivalent to adding white, and the resulting color is less saturated than the spectral cyan. The only way to match spectral cyan with a mixture of B and G, therefore, is to add red to cyan and then match it.

Adding red to the spectral cyan means to desaturate the cyan. Since red is the complementary color to cyan, the addition of red is equivalent to adding some white and produces a less saturated color. Adding red to the spectral cyan we were trying to match is equivalent to adding negative red to the mixed color. This is physically impossible: Since there was no red to begin with in the B + G mixture, it is not possible to subtract red from it.

In other words, we cannot add negative R to G + B, but we can, equivalently, add positive R to cyan, as we can see from the simple algebra:

$$\text{saturated } C = G + B - \varepsilon R$$

where εR designates an appropriate amount of R (the amount of red required depends on the particular RGB primaries used). Moving $-\varepsilon R$ to the left side of the equation:

$$\text{saturated } C + \varepsilon R = G + B$$

In fact, this is how color matching of spectral colors with RGB is done: Adding red light to the saturated cyan and adjusting the amount of R until the cyan matches the G + B.

A negative amount of R and, consequently, a negative fraction of red r can indeed be plotted on the color triangle or, rather, outside the color triangle, as shown in Figure 7.14.

The spectral cyan lies outside the color triangle. The color triangle defines the colors that can be matched by a mixture of RGB, but more colors do exist—they just can't be matched by RGB. Cyan is an example of these colors. Many other spectral and non-spectral colors lie outside the triangle. These colors have to be desaturated before they can be matched by a mixture of RGB primaries. You will never see an RGB-mixed turquoise on TV or on your computer monitor that is as saturated as the spectral turquoise.

The hues obtained mixing primaries are less saturated than their spectral corresponding colors, which are each associated with a single (dominant) wavelength.

As shown in Figure 7.15, most of *the spectral colors lie outside the color triangle*. Spectral red, green, and blue are at the edges of the triangle, while yellow, cyan, and violet are outside of it. So are the high-saturation purples.

The horseshoe shape in Figure 7.15 represents *the area of human color vision*. This area is enclosed by the curved line along which all spectral colors

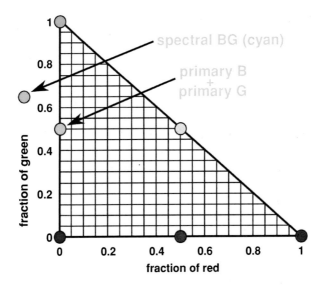

FIGURE 7.14 ▶ The difference between spectral C and the C obtained mixing B and G is evident on the color triangle: spectral cyan requires negative red, therefore it lies outside the color triangle.

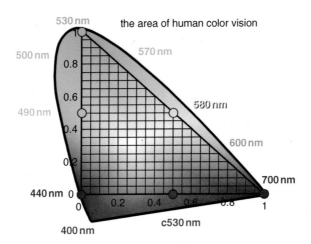

FIGURE 7.15 ▶ The spectral colors, which we can see but we cannot mix using RGB, lie outside the color triangle. The colored horse-shoe shape in this figure contains the area of human color vision, while the triangle contains only the colors that we can mix using RGB. Notice how much smaller the area of the triangle is, compared with the area of human color vision.

lie (see corresponding wavelengths in the diagram) and the line between spectral violet and red. In summary, *the color triangle includes only the colors that can be obtained mixing RGB, but the eye can see more colors, including spectral colors and purples*.

7.6   Non-Spectral Colors

Purple and magenta, as well as all the magenta-purple hues, are *non-spectral colors*. There is no wavelength associated with them because they are not part of the spectrum of white light. They are exclusively mixed colors. Magenta is the mixture of spectral red (700 nm) and spectral blue (440 nm) or spectral violet (400 nm), while all purples are simply low-brightness magentas. The mixture of R + B in different proportions generates all the magentas: the reddish magentas and the bluish magentas. These hues can be obtained mixing RGB primaries (R + B), and therefore you can see them on TV or on a computer monitor. The magenta and purple hues obtained mixing violet and red cannot be mixed with RGB primaries, since they lie outside the color triangle.

Now that we described the color triangle and the spectral colors, let us look at the complementary colors again. Two additive complementary hues must produce white when added together. White must be on the line between the two colors. For example, B and Y are complements because when additively mixed they produce white. To be perfectly accurate, B and yellow are complementary hues, not complementary colors. B and spectral Y are not complementary colors, since the points B and Y are not equally distant from white, while SY and B are equidistant from W. We would actually need to add a lower saturation blue (sky blue, labeled SB in Figure 7.16) to yellow if

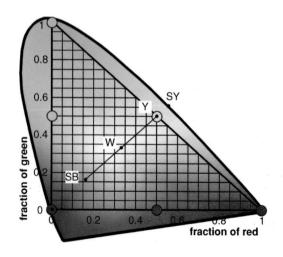

FIGURE 7.16 ▼
The true complementary color to blue (B) is spectral yellow (SY). The complementary to Y is sky blue (SB). SY and B are equidistant from white (W), as are Y and SB, therefore, added in equal amounts they produce white.

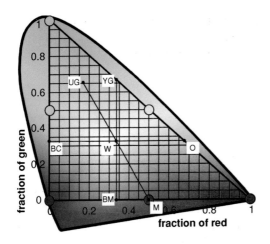

FIGURE 7.17 ▼
Other pairs of complementary colors are: bluish cyan and orange (BC and O), lime green and bluish magenta (YG and BM), unsaturated green and magenta (UG and M).

we wanted to obtain white. As an aside, note that B and Y pigments do not produce white when mixed, as will be explained in detail in Chapter 8.

The line connecting blue and yellow passes through white. Other pairs of complements are shown in Figure 7.17, but many more exist. Again, to be accurate, it is unsaturated green (UG) that complements magenta perfectly.

We can now explain how purples and magentas can be defined by a single wavelength, as described in Chapter 6 in the definition of *hue*. To do so, one must illuminate magenta with light of appropriate intensity and wavelength until it becomes white. The wavelength and saturation of the color used are the complementary color to magenta—that is, green. Magenta is, therefore, accurately defined by its complement, as indicated in Figure 7.15: c530 nm, where c stands for complementary.

7.7 Summary

Every color can be located on the color triangle by a three-step process. First match the color with three primaries R, G, and B. Second, find r and g:

r = fraction of red in the total intensity of light in the match

g = fraction of green in the total intensity of light in the match

Third, plot r on the horizontal axis and g on the vertical axis.

White falls at the point r = 0.33, g = 0.33, near the center of the triangle. The more saturated a color (i.e., the greater its purity), the further it falls from white. Spectral colors are the most saturated and fall on a curved line outside the triangle.

When two colors are added, the resulting color always falls along the line connecting the two. This idea is useful in the explanation of complementary colors—colors that produce white when added. Complements always fall on opposite ends of a line through the white point, The hue of a color can be specified by finding the dominant wavelength or, in the case of purples, the complementary wavelength of the color. Every color can be matched by a spectral color, or a purple, plus white. The smaller the amount of white, the higher the saturation.

7.8 Additive Color Mixing in Painting

When painting and mixing oil paints, the rules of additive color mixing don't usually apply. We will describe in Chapter 8 how to mix pigments and paints with subtractive color mixing. Nevertheless, there have been a few extraordinary cases in art in which painters have attempted to paint additively.

Postimpressionists used a revolutionary technique to enhance color brightness and contrast. This technique was based on additive color mixing. Van Gogh wanted to achieve the effects of the art glass in the windows of Gothic cathedrals. These windows have colored pieces of glass adjacent to each other, and they transmit colored light exactly as projectors do through colored filters.

Van Gogh did not know about additive color mixing and mixing lights, but he was striving to achieve the brightness and color contrasts that only mixing lights can achieve. He invented the *mélange solide,* French for "solid mixture," in which instead of mixing oil colors subtractively, he painted adjacent brushstrokes with contrasting colors (see, for example, *Self-Portrait with Hay Hat,* by Vincent Van Gogh [Dutch, 1853–1890], Postimpressionist). Van Gogh said of his painting, "Je cherche à exprimer avec le rouge et le vert les terribles passions humaines" ("I try to express with red and green the terrible human passions"). Since the brushstrokes in his *mélange solide* are large—much larger than the eye's resolution—the eye does not actually mix lights from adjacent colors, but it perceives the striking contrast of very different and very bright colors, as in a Gothic cathedral window!

If the regions of two different colors are much smaller, lights from the two different colors side by side are mixed additively by our eyes and create a third color. This resulting color is described as optically mixed—that is, it is mixed in the eye of the observer. When two colors are mixed with pigments by a painter (subtractively), they lose some of their intensity, while when the same two colors are mixed optically, they retain their intensity and they may appear brighter. Two Neo-Impressionists, the *pointillists* Seurat and Signac, explored this possibility. They experimented with small color regions: points or dots, as shown in their paintings. See, for example, *Chenal de Gravelines*, by Georges Seurat (French: 1859–1891), and *St. Tropez, le Port,* by Paul Signac (French: 1863–1935).

Seurat studied the effects of different colors interacting with each other, and he tried to control their mixing by reducing the number of colors on his palette and applying pigments in very small spots. The Impressionists have often been interpreted as using "broken colors," but a patch of yellow in an impressionist painting is almost always yellow paint, not strokes of red and green laid side by side to achieve optical mixing. Seurat, however, did construct magenta from small spots of blue and red, and with his system, he could also vary the intensity of red—for example, by placing it near an orange dot and determining how much the red will induce a perception of its complement (cyan) in the adjacent color.

Seurat called his technique divisionism because he had divided the colors into their physical elements. His theory and practice, together with those of his followers, were soon described as Neo-Impressionism and Pointillism. The subject was still Impressionist—ordinary people seen in every day activities, atmospheric conditions, immediacy of the scene—but the presentation and technique were new.

The Neo-Impressionists had color theory available to them, including the works of Herman von Helmholtz (whose *Manual of Psychological Optics* had by then appeared in French) and Wilhelm von Bezold (whose texts were published in French in 1876). Seurat and his colleagues had begun to feel that if they did not become more involved in the science behind their technique, they would not be able to explain the optical effects forming the basis of visual art and art itself would remain intellectually unsatisfying. Seurat's task was made easier by the work of the American physicist Nicholas Odgen Rood (see *Modern Chromatics with Application to Art and Industry*, Chicago 1879), who was himself an accomplished painter. Seurat was inspired by Rood's writings and color theory, and he imitated the behavior of light itself, performing additive color mixing when actually optically mixing pigments. However, he was incorrect in thinking that "adjacent dots of pure spectral colors could be mixed [in the eye of the observer] to obtain other spectral colors observed in nature, with a brilliance and luminosity [read saturation!] unattainable by conventional pigment mixtures." Scientifically speaking, Seurat's idea was flawed. Nevertheless, he has given us some of the most haunting images of European art. Another problem was with the contrast enhancement of adjacent complementary colors. We have now learned that lights of complementary colors produce white when mixed—*not* saturated colors.

The optical mixture of complementary colors kills color, as Signac himself later admitted: "Pointillage makes the surface more lively, but it does not guarantee luminosity, intensity of color, or harmony. The complementary colors, which are allies and enhance each other when juxtaposed, are enemies and destroy each other if mixed, even optically. A yellow and a blue, if juxtaposed, enliven each other, but yellow dots and blue dots make an aggregate which is grey and colorless." You can verify this effect very simply. Next time you go to the Art Institute of Chicago, stand in front of *Dimanche après Midi sur l'Isle de la Grande Jatte*, by Georges Seurat. If you are close enough to see the individual dots, all colors appear very bright and luminous (saturated). If you stand far away they all blend together, and the overall effect is a hazy gray patina covering the whole painting.

More recently, a contemporary artist, Chuck Close (American, b. 1940) was also inspired by optical mixing of colors. His portraits and self-portraits are subdivided into a grid of large squares, each of which is an abstract painting containing several complementary, contrasting, or similar colors. Looking at Close's paintings or prints closely reveals all these colors, while standing far away from them blends them all together, optically. An example is shown in Figure 7.18.

FIGURE 7.18 ▼
Lyle, 2003. Chuck Close, silkscreen, 149 colors, 166cm × 137cm. Edition 80. The 1149 squares perfectly blend together to form the portrait of photographer and performer Lyle Anton Harris. Reprinted with permission from Dick Solomon of Pace Prints.

7.9   Questions

1. At a hip-hop concert the band is illuminated by red and blue spot lights. What color does the white T-shirt of the DJ appear?

2. Identify on the color triangle eight colors of your choice. With the help of the color triangle, write down the amount of R, G, and B corresponding to each color. Convert the amounts you obtained into the 8-bit graphics numbers so that the amounts of R, G, and B range between 0 and 255. If you have a computer, also create a graphic file with eight separate fields and fill each one of them with the colors you have just identified.

3. Add light of a color of your choice to another color of your choice, and identify the resulting color on the color triangle. Choose mixed colors, not pure R, G, or B.

4. Show where the complements of the marked colors are on the triangle. What colors are they? Hint: First locate the "white point."

5. What hue is each of the following combinations of primary colors?

   (a) $0.3R + 0.1G + 0.6B = ?$
   (b) $0.2R + 0.2G + 0.2B = ?$
   (c) $0.1R + 0.5G + 0.4B = ?$
   (d) $0.2R + 0.2G + 0.1B = ?$

6. Most people think that the primary colors are blue, red, and yellow, instead of green. If you chose as primary colors B, R, and Y, what hues could you reproduce by adding these three primaries? What hues could you not reproduce (e.g., could you make the purples, cyan, orange, white)?

Chapter 8

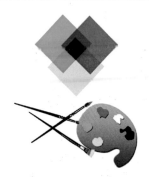

Subtractive
Color Mixing

Subtractive color mixing applies to overlapping colored filters and color photography and is essential for understanding the mixing of pigments in paints or inks and for color printing.

8.1 | Filters

If light of a given spectral distribution passes through a *filter*, the spectrum of the light is modified. The filter absorbs part of the light and transmits the remainder.

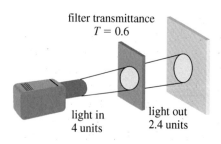

filter transmittance
$T = 0.6$

light in
4 units

light out
2.4 units

FIGURE 8.1 ▶ A neutral density filter that transmits 60% of the light has transmittance $T = 0.6$.

The *transmittance T* of a filter is the fraction of the incident light transmitted through the filter. The filter in Figure 8.1 has $T = 0.6$. The filter absorbs 40% and transmits 60% of the light. If the incoming light has an intensity of 4 units, then the transmitted light has an intensity of $4 \times 0.6 = 2.4$ units.

The transmittance of a filter is usually different for different wavelengths (colors) of light. The transmittance is only the same for all

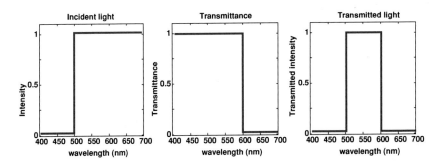

FIGURE 8.2 ▶   Idealized spectra of colored light incident on a colored filter. The spectrum of the illuminating light is on the left, the transmittance curve of the filter at the center, and the spectrum of the transmitted light on the right. You should be able to decide what color the incident and transmitted lights are from these plots.

wavelengths if the filter is a *gray filter,* also called a *neutral density filter.* Let us look at a colored filter in Figure 8.2.

Once again we are using a simplified version of the spectrum, in which the range of wavelengths between 400 and 500 nm is referred to as blue, 500–600 nm is green, and 600–700 nm is red, respectively (B, G, and R). The real spectra are curved lines but they can be simplified as step functions, with vertical and horizontal lines, as in Figure 8.2. While this approximation is not *always* perfect to describe the behavior of light and color, it works in most cases, and it is much simpler to use than the real spectral lineshapes. Examples of real filter curves are reported in Figure 8.3.

If you were not able to correctly identify the colors in the spectra of Figure 8.2, here is a complete explanation. The first spectrum was the incident light. This light has no intensity (Intensity = 0) in the B region of the spectrum and intensity $I = 1$ in both the G and R regions. The incident light is therefore a mixture of G + R lights. As described in Chapter 7, G and R lights combine to form Y light. Therefore, you know that the incident light is yellow. The filter transmits all light in the B and G regions (where Transmittance = 1) but does not transmit light in the R region (where Transmittance = 0). Since there was no B in the incident light, the only light that the filter transmits is G. Therefore, the spectrum of the transmitted light has Intensity = 0 in both the B and R regions and Intensity = 1 in the G region, transmitting all the green light and causing the transmitted color to be green. The filter *subtracted* red from the incident light. It *absorbed* the R component from the incident light. We are now beginning to see why this is called *subtractive color mixing*.

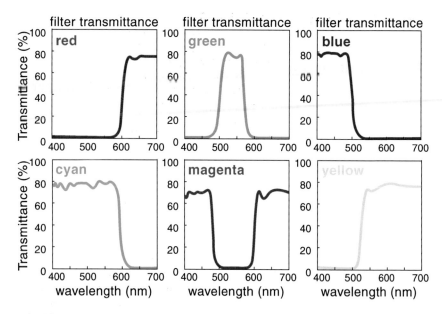

FIGURE 8.3 ▶ Transmittance curves of real filters. These are R, G, B interference filters (top) and C, M, Y interference filters (bottom). The filters used are commercial dichroic filters. Reprinted with permission from P.U.P.A. Gilbert and W. Haeberli, American Journal of Physics, **75,** 313–319 (2007). Copyright © 2007, American Association of Physics Teachers.

8.2 | Subtractive Primary Colors

The most commonly used *subtractive primary colors are cyan, magenta, and yellow.* Each one of these colors absorbs one additive primary, its complementary color.

Cyan absorbs red, magenta absorbs green, and yellow absorbs blue. When combined, *C, M, and Y produce black.* Again, these are not the only colors that can be chosen as subtractive primaries. Any other three colors that combined produce black are potential primaries, although the mixing potential—the number of colors that can be obtained mixing three primaries—is maximum with CMY.

This is why the computer industry has chosen CMY inks to be used in color printers, and the same choice was made by all manufacturers for color photography.

Figure 8.4 shows the absorption of the three subtractive primaries when CMY filters are used. Figure 8.5 shows how to obtain the additive primaries using combinations of CMY filters. In general, paints, inks, and pigments

Subtractive primaries

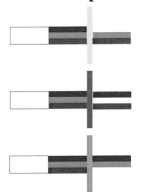

FIGURE 8.4 ▼
White light, coming from the left, for simplicity's sake, is considered to have only RGB components. *Each filter subtracts its complementary color*. Each filter (or pigment) absorbs its complementary color and transmits (and partly scatters) the other colors.

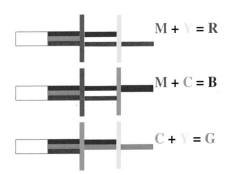

$$M + Y = R$$

$$M + C = B$$

$$C + Y = G$$

FIGURE 8.5 ▼
Combinations of subtractive primaries to obtain R, G, and B. The order of the two filters is not relevant: Swapping them will produce the same color.

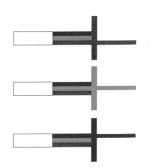

FIGURE 8.6 ▼
R, G and B filters transmit the R, G, and B light. They absorb their complements just as Y, M, and C filters do: Red absorbs B and G (C), green absorbs B and R (M), blue absorbs G and R (Y), as shown.

behave the same way with scattered light as filters do with transmitted light. Figure 8.6 shows the absorption of R, G and B filters, which also absorb their complementary colors (see Figure 6.14 for the definition of *complementaries*).

8.2.1 Subtractive primaries

In subtractive color mixing, one often obtains colors that seem unexpected or surprising. The *hue* **and the** *saturation* of the light transmitted through two consecutive filters is not intermediary between the hues of the two filters. The same applies for pigments. You can look on the color triangle to see where the colors lie and compare that with the colors obtained anywhere in this chapter when overlapping filters or mixing pigments.

Combinations of overlapping filters produce the results shown in Figure 8.7.

Notice from Figure 8.7 that the overlapping of red (or M + Y) and green (or C + Y) filters is black. Therefore,

$$\text{Subtractively}, R + G = \text{Black}$$
$$\text{Additively}, R + G = \text{Yellow}$$

You can see that subtractive color mixing is very different from additive color mixing! Mixing filters (or pigments and paints) is very different from mixing lights.

As discussed, overlapping filters and mixing pigments, paints, or inks are called *subtractive color mixing*. The term is slightly misleading because a filter does not subtract (absorb) a fixed amount of light, but, rather, it absorbs a fixed *fraction* of the light incident upon it. The mathematical operation is not a subtraction. It is the **multiplication** of the incident intensity by the transmittance of the filter or the multiplication of the transmittances of two filters.

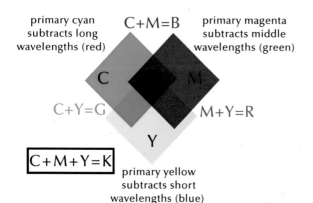

primary cyan subtracts long wavelengths (red)

C+M=B

primary magenta subtracts middle wavelengths (green)

C

M

C+Y=G

M+Y=R

Y

C+M+Y=K

primary yellow subtracts short wavelengths (blue)

FIGURE 8.7 ▼
Three overlapping C, M and Y filters. Each two of these combined transmit an additive primary color (R, G or B), while all three do not transmit any light (K = black).

Let us look more in detail at yellow and cyan filters, overlapping to transmit green light, as shown in Figure 8.8.

Figure 8.9 shows the transmittance curves and table for this example.

The resulting transmittance of the two filters can be graphically obtained by combining the transmittance curves of the two filters. In Figure 8.10 you can see that the region in which the two curves overlap (the dashed region) is the resulting transmittance. In this region neither of the filters had a transmittance of 0.

FIGURE 8.8 ▶ White light, coming from the left, can be for simplicity's sake considered to have only RGB components. *Each filter subtracts its complementary color.* The cyan filter absorbs R (its complementary color) while transmitting B and G. Remember that additively B + G = C. Therefore, the B and G lights transmitted by the filter form C. The yellow filter absorbs B and transmits G and R. Again, additively G + R = Y, so a C filter transmits C light, a Y filter Y light. However, the combination of the two filters will only transmit G light. If we start from white light, we have R, G, and B. The cyan filter subtracts R and the yellow filter subtracts B, leaving only G to be transmitted. (R + G + B) − R − B = G.

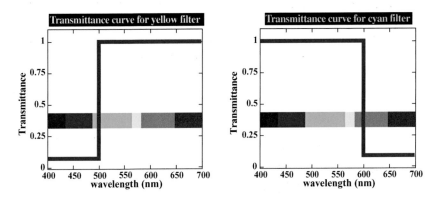

FIGURE 8.9 ▶ Transmittance curves for Y and C filters.

Simplified spectral ranges	T_{yellow}	T_{cyan}	$T_{Y\&C} = T_Y \times T_C$
B (400–500 nm)	0	1	0
G (500–600 nm)	1	1	1
R (600–700 nm)	1	0	0

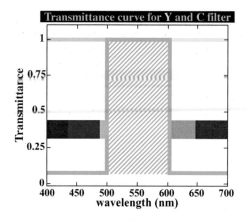

FIGURE 8.10 ▼
Light transmitted through Y
and C filters, as in Figure 8.9, is
green. The only region in which
both C and Y spectra are non-
zero is the green region. If you
prefer to think graphically: The
only region in which the two
spectra of Figure 8.9 overlap is
the dashed region, which corresponds to green.

8.3 Color Photography

All color films consist of three layers containing silver halide (silver iodide or silver bromide) emulsion, coated on top of each other to form a permanent, multilayered structure supported by the film base. Each emulsion layer of color negative film absorbs a specific color, determined by the properties of the dyes embedded in that layer. There is a chemical reaction in the emulsion that permanently modifies the layer after it is exposed to and has absorbed light of a specific color. The top emulsion layer absorbs B light and transmits G and R light to the lower layers. The middle emulsion layer responds only to G, and the lower responds only to R.

Objects with colors other than B, G, or R, which illuminate the film during exposure, react with two or more of the emulsion layers, according to the spectral distribution of the light they scatter. The original color of the object is reproduced only when the negative is printed. *Color negative film* produces an image that is complementary in respect to colors and density of the original objects. Subsequently, the negative is printed on color photographic paper to yield *a positive color print*.

After exposure, the negative film is developed to yield the color negative. The processing steps include immersion in a color developer, a bleach rinse, a fixer, a bleach/fix, a final wash, a stabilizer, and drying. Let us briefly look at these steps separately.

The chromogenic developer converts light-sensitive crystallites of silver halide compounds in the emulsion layers into metallic silver. While doing this, the developer also oxidizes and combines with dye couplers that are either built into the emulsion layers or added during development. The result is the formation of three dye layers, one from each of the subtractive primary colors: cyan, magenta, and yellow. The blue-sensitive layer of the original film forms a yellow image, while the green- and red-sensitive layers form magenta and

| unexposed color film (side view) | exposed and developed color film | bleached and fixed color film |

FIGURE 8.11 ▶ The unexposed film (left) appears gray. Once it has been exposed and developed the B, G, and R absorbing layers are colored. After bleaching and fixing, the silver halide crystals are removed, and only the color layers remain but only in the appropriate places to produce a negative of the image. Where only the cyan layer is present, the image will be printed red, and so on.

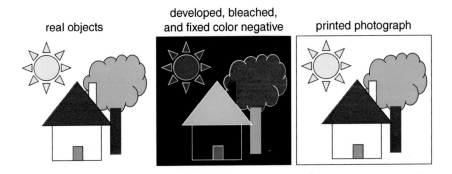

real objects | developed, bleached, and fixed color negative | printed photograph

FIGURE 8.12 ▶ The processed color film is a negative image in which **all colors are inverted**—that is, the colors are complementary to colors of the original objects. These colors, such as the cyan roof and the blue sun, can be seen in transmission on the negative and act exactly like the combination of subsequent CMY filters. In the negative, the C layer is only present where the roof is located in the image and the M and C layers are only present where the sun is. The Y layer is absent from both areas of the processed negative. The black areas on the negative have all three layers. White light is then transmitted through the processed film and projected onto color-sensitive paper for printing. The paper is coated with a similar emulsion that inverts the colors all over again!

cyan images, respectively. Bleach is then utilized to remove all of the silver metal so that each layer is left with only a color image. After bleaching, the film is fixed and washed thoroughly to remove all dissolved silver salts. Then it is rinsed in stabilizer to improve dye stability and to harden the emulsion. Finally, it is dried. Figures 8.11 and 8.12 illustrate the use of color layers and film processing.

8.4 Pigments

The color of familiar objects depends on the **pigments** that they contain and the **reflectance R** of these pigments. Consider the leaf of a tree, illuminated by

Simplified spectral ranges	Incident intensity	R_{leaf}	Reflected light
B (400–500 nm)	0.8	0	0
G (500–600 nm)	0.9	1	0.9
R (600–700 nm)	1	0.5	0.5

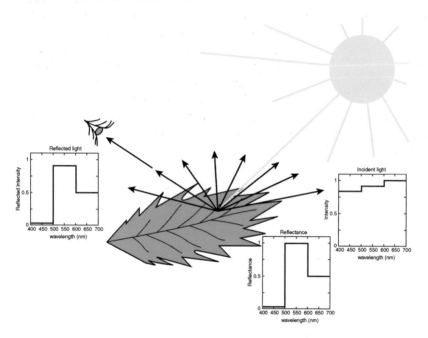

FIGURE 8.13 ▶   Sunlight illuminating a green leaf. The simplified spectra for the incident light, the reflectance, and the reflected light are also shown.

sunshine. Some of the light from the sun is specularly reflected as unchanged white light by the glossy surface of the leaf. The remainder of the light is partly absorbed and partly scattered in all directions, as shown in Figure 8.13. The reflectance is the fraction of the incident light that is scattered. As in the case of transmittance, the reflectance usually depends on the wavelength of light. The reflectance indicates the amount of light scattered by the leaf in each of the separate RGB ranges, as shown in the table in Figure 8.13. The reflectance determines the color of the leaf, as we see it.

Pigments are the colored particles suspended in paints, inks, color photographic film, and so forth. When two paints of different colors are mixed, a third color is obtained, which is different from each of the two components. Can we anticipate what color will result? Artists do this regularly, using their "experience"—that is, after preparing and seeing many mixtures of different colors, they can "remember" the resulting colors. Here we analyze the physical reason for obtaining the color of the mixture. Pigments act very nearly as filters and are mostly subtractive. The reason pigments in a mixture "subtract" color from each other is called ***multiple scattering***. Let us explain this concept with an example: the mixture of red and blue paints. Red pigments absorb all blue and green from the illuminating white light and diffusely reflect only red light. The reflected red light is in turn almost completely absorbed by nearby blue pigments. A similar but inverse behavior is observed at all blue pigments. Red and blue pigments, therefore, absorb each other's scattered light. This phenomenon happens not only once but many times in the paint mixture, which is why we said the phenomenon is called *multiple* scattering. There is a simple experiment that proved how *multiple* this multiple scattering is. Adding a small amount of black to paint of any other color has a large effect (darkening) on the color of the mixture. See how striking this effect is in the experimental data of Figure 8.14.

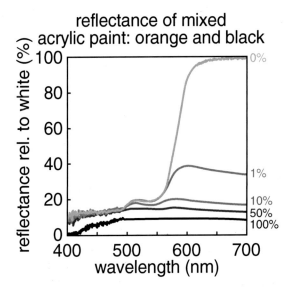

FIGURE 8.14 ▶ The reflectance curves of orange acrylic paint, black acrylic paint, and mixtures of the two, in different proportions. The mixed volumes were measured using Gilson pipettes and the percentage of black paint is reported for each curve by the right axis. The colors used to plot each spectrum are representative of the color of paint or paint mixture. Notice how a small amount of black paint (1%) reduces the intensity to 40%. This result strongly suggests multiple scattering, that is, each black pigment has many chances to absorb light, not just one. Reprinted with permission from P.U.P.A. Gilbert and W. Haeberli, American Journal of Physics, **75**, 313–319 (2007). Copyright © 2007, American Association of Physics Teachers.

Quantitatively, when mixing pigments of two different colors, the resulting color depends on the reflectance curves for the two colors. The reflectance values in each spectral range can be calculated as just described for the transmittances of filters.

It in worth noticing that the reflectance values in each spectral range correspond exactly to the RGB intensities we discussed in additive color mixing, and all colors, when defined *only* by their reflectance or transmittance curves, can be located on the color triangle. All you have to do is use the curves to calculate the fraction of red r and the fraction of green g!

On a computer, magenta and cyan would have the RGB intensities:

$$M = 255B + 0G + 255R$$

and

$$C = 255B + 255G + 0R$$

We already discussed the fact that the units and the absolute values of each color intensity (a, b, and c) are not relevant. The only relevant quantities are the fractions of red and green.

The fraction of red and green for M and C are

$$r_M = 0.5 \text{ and } g_M = 0$$

$$r_C = 0 \text{ and } g_C = 0.5.$$

Based on these coordinates, it is obvious where M and C are positioned on the color triangle. When subtractively combining colors of known reflectances, if the resulting reflectance is a difficult color to identify, the fractions of red and green can easily be calculated and the color can be identified on the color triangle.

With this new tool, let us now look at a few examples of subtractive color mixing that show changes in saturation or hue.

8.5   Change in Saturation

Sky B	+ Orange	= Orange (pure, low intensity)
B = 255	B = 0	B = 0
G = 128	G = 128	G = 64
R = 128	R = 255	R = 128

From the RGB intensities given for each color, we can draw the reflectance curves and fill in the reflectance table.

Simplified spectral ranges	R_{uB}	R_O	$R_{uB\&O} = R_{uB} \times R_O$
B (400–500 nm)	1	0	0
G (500–600 nm)	0.5	0.5	0.25
R (600–700 nm)	0.5	1	0.5

Sky blue is unsaturated B (uB in the table)—that is, B + white. The B reflectance curve therefore has white in it—in other words, a certain amount of all three spectral components, R, G, and B (0.5 in this example), shown in Figure 8.15. Adding white is equivalent to adding a horizontal line in the reflectance curve that shifts the whole reflectance curve up. Orange, as should be evident from the RGB components and from the reflectance curve of Figure 8.15, is saturated. *As long as at least one of the RGB components is 0, the color is saturated.* As long as the reflectance curve has an intensity of 0 in at least one spectral range, there is no horizontal line added to the curve—there is no white and the color has high saturation.

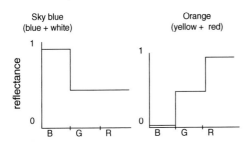

FIGURE 8.15 ▶ The reflectance curves for sky blue and orange.

The reflectance curve for orange is a direct consequence of how orange is additively obtained: 1 part of Y light and 1 part of R light. But Y = R + G lights, so orange must contain 2 parts of R and 1 part of G, as shown in the reflectance curve in Figure 8.15.

The color resulting from the mixing of sky blue and orange is saturated, since its reflectivity in the blue range is 0. The reflectance curve for this color is graphed in Figure 8.16. This curve is identical to the one for orange except for a lower intensity; this color is low-brightness orange, or brown.

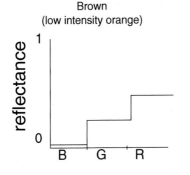

FIGURE 8.16 ▶ The resulting color is saturated brown, or low intensity orange.

8.6   Why Do Blue and Yellow Make Green?

Have you ever tried to mix blue and yellow paints? If you did, you certainly obtained green. Now, let us look at the idealized reflectance curves for blue and yellow paints, shown in Figure 8.17. There is no spectral region in which the reflectance curves for yellow and blue overlap. In other words, at every wavelength the product of the reflectances is zero. Since this means no light in the 400–700 nm region is reflected and all is absorbed, the color produced should be black. So Figure 8.17 must be wrong. In fact, its spectra are over-simplified. The spectra of light scattered by yellow and blue paints do not have such sharp cutoff but rather look like the ones in Figure 8.18.

Remember that when we introduced the "oversimplified" spectrum in Chapter 6, composed by only the three colors RGB, we said that it works in *most* cases? Well, this is one of the cases in which that simplification does *not* work. We cannot consider the reflectances of yellow and blue as having three values only (0, 1, 1 for yellow and 1, 0, 0 for blue, in the B, G, and R regions, respectively). We must consider that the real reflectance curves have many more values, not only three.

When measuring the reflectance of the resulting paint color, we find that it is nonzero in the green region, as shown in Figure 8.18. That is why the mixture of blue and yellow is green.

Whenever the simple answer to the question "What color will we obtain if we mix two colors?" is "black," one must be cautious and not trust the oversimplified spectrum with three colors only. There may, in these cases, still be significant overlap of the real reflectance curves of the two colors. In all other cases of color mixing we discussed before and will discuss later, the oversimplification of considering only three colors (RGB) in each reflectance curve does not constitute a problem, and the results can be trusted.

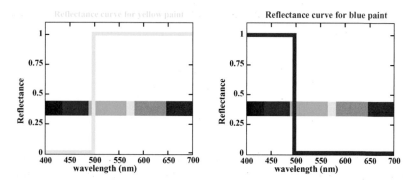

FIGURE 8.17 ▶ Oversimplified Y and B reflectance curves lead to the wrong conclusion that yellow and blue make black.

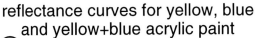

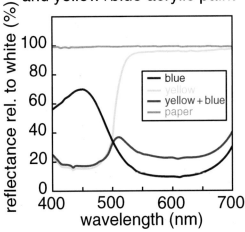

FIGURE 8.18 ► The measured reflectance curves for yellow, blue, and yellow + blue acrylic paints. Notice that, since yellow and blue are complementary colors, the resulting mixed color should be black, while in reality it is a dark green, with a maximum in the only region in which both spectra are non-zero—that is, green. The color used to plot each spectrum is representative of the color of paint or paint mixture. Reprinted with permission from P.U.P.A. Gilbert and W. Haeberli, American Journal of Physics, **75**, 313–319 (2007). Copyright © 2007, American Association of Physics Teachers.

8.7 Change in Hue

We also mentioned that in most cases, mixing pigments or overlapping filters is equivalent. There are a few exceptions to this rule as well. The following two examples only apply to filters. Mixing two parts of the same color with pigments (in paints or inks) generates twice the amount of the same color, and there is no change in hue.

$$\textbf{M-purple} + \textbf{M-purple} = \textbf{BluishM}$$

M-purple	M-purple	BluishM
B = 255	B = 255	B = 255
G = 0	G = 0	G = 0
R = 128	R = 128	R = 64

However, the light transmitted by two subsequent, identical magenta-purple filters is not magenta-purple but bluish magenta! Look at the transmittance table for an explanation of this unexpected result and the reflectance curves for the corresponding colors in Figures 8.19 and 8.20.

Simpl. ranges	T_{MP}	T_{MP}	$T_{BM} = T_{MP} \times T_{MP}$
B	1	1	1
G	0	0	0
R	0.5	0.5	0.25

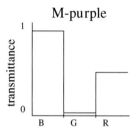

M-purple

FIGURE 8.19 ▶ The transmittance curve for magenta-purple: 2 parts of B, 1 part of R.

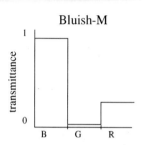

Bluish-M

FIGURE 8.20 ▶ The resulting color, transmitted through two identical magenta-purple filters is not magenta-purple, but bluish-magenta.

Two orange filters **when used in combination transmit** reddish-orange

FIGURE 8.21 ▼
The color transmitted by two identical orange filters is not orange, but reddish orange.

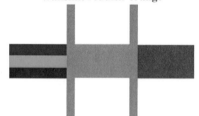

The same phenomenon occurs when using two orange filters, as shown in Figures 8.21 and 8.22. The resulting color is then identified on the color triangle in Figure 8.23.

After the first filter

$$r = \frac{1}{1.5} = 0.7 \quad \text{and} \quad g = \frac{0.5}{1.5} = 0.3$$

After the second filter

$$r = \frac{1}{1.25} = 0.8 \quad \text{and} \quad g = \frac{0.25}{1.25} = 0.2$$

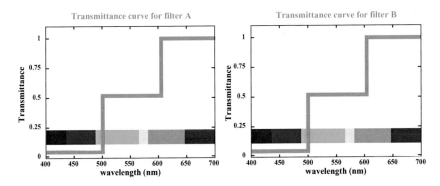

FIGURE 8.22 ▶   The transmittance curves of two identical orange filters.

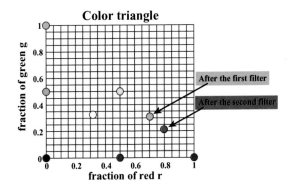

FIGURE 8.23 ▼
The color transmitted by the two filters in Figure 8.22. After the first filter the color is orange, but after the second filter it becomes reddish-orange.

Using the values of r and g, we can locate the two hues on the color triangle, as shown in Figure 8.23.

Simpl. ranges	T_O	T_O	$T_{RO} = T_O \times T_O$
B	0	0	0
G	0.5	0.5	0.25
R	1	1	1

Although these surprising changes in hue do not really apply to mixing pigments, painters are well aware of them. This effect occurs to a limited extent in oil or acrylic painting, but it affects watercolor painting much more significantly. As painters know, one layer of watercolor on white paper is partly transparent and, therefore, acts as a colored filter in front of a projector. If the painter paints a second layer over the first with the exact same watercolor, the hue changes slightly!

8.8   Questions

1. The transmittance curve of a filter tells you what fraction of the light of a particular wavelength the filter transmits. A transmittance of 1 in the following graphs means all the light of that wavelength is transmitted through the filter; 0.5 means half of the light passes; and 0 means no light goes through at that wavelength. What color is the light from filter A? From filter B?

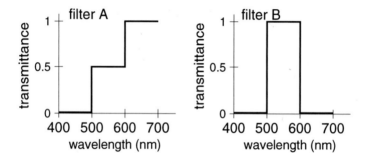

2. **Subtractive mixing.** What color do you see on the screen if you use filters A and B defined in problem 1 in the following configuration (left)?

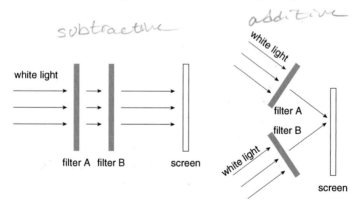

3. **Additive mixing:** What color do you see on the screen if you use filters A and B defined in problem 1 in the preceding configuration (right)?

4. Answer problems 1, 2, and 3 using filters C and D, described below.

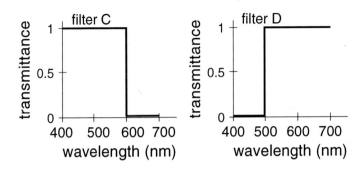

5. Answer problems 1, 2, and 3 using filters E and F.

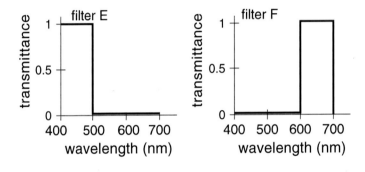

6. Answer problems 1, 2, and 3 using filters G and H.

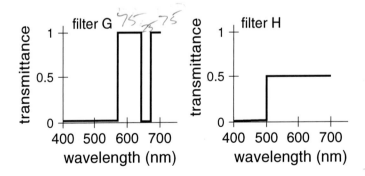

7. What color do you obtain if you add two parts of C paint and one part of Y paint? Find the color reflectance curve first, and then report the coordinates of this color on the color triangle, and see what color this is.

Chapter 9

Color-Generating Mechanisms

9.1 | Illuminating Light

The color we see depends on the illuminating light. When an apple is illuminated by white light, it appears red, but when it is illuminated by only blue light, it appears black. This is because the pigments in the apple absorb green and blue light and scatter red light. When there is no red light to be reflected, all the blue or green light is absorbed, and the apple does not reflect light of any wavelength. This is why it appears black.

Pieces of paper of assorted colors when illuminated by yellow light, such as a sodium lamp, will only appear colored if their pigments reflect a certain amount of yellow wavelengths, that is, if they are greens or reds or whites. Any blue shades will appear black, as they completely absorb yellow light.

9.2 | Pigments

The colors of the familiar objects surrounding us most frequently depend on the pigments in the objects. Different pigments absorb and scatter light of different wavelengths and therefore appear colored themselves. There are animal, plant, and mineral natural pigments, as well as synthetic dyes and coloring agents for paints, fabrics, food, and so on. The animal pigments include:

- **melanin (which can be black, gray, beige, or brown)**
- **carotenoids (which can be yellow, orange, pink, or red)**

Physics in the Arts, Revised Edition

FIGURE 9.1 ▶ Fall colors at the UW-Arboretum in Madison, Wisconsin.

Melanin gives color to our skin and hair, with the exception of red hair, which has a very unusual iron-oxide pigment. On the other hand, carotenoids give color to the beaks and feet of birds; another carotenoid (astaxanthin) colors the shells of crabs, shrimp, prawns, and lobsters and becomes the dominant color when these animals are cooked. There are also blue animal pigments, present only in fish and shellfish. The blue eyes in humans and Siamese cats are *not* due to pigments but to scattering.

The most common plant pigments are:

- **chlorophylls (which is green)**
- **carotenoids (which can be yellow, orange, pink, or red)**
- **tannin (which is black or brown)**
- **anthocyanins (which can be red or bluish purple)**

9.3 Structural Color: Iridescence

Some birds, such as parakeets, kingfishers, peacocks, hummingbirds and mallard ducks, have bright blue or green feathers. The shells of many beetles

are also green. The wings of the Brazilian butterfly *Morpho peleides* are a brilliant blue-cyan (Figure 9.2). But with no blue or green pigments in animals, what gives these animals their striking colors?

These colors are generated by *iridescence*. Iridescence is an optical phenomenon separating the spectral colors present in white light. Iridescent colors are generated by the structure of objects, not by the presence of colored pigments. In this sense they are *structural colors*. If a material is subdivided or layered very finely—so finely that the spacing between subdivisions or layers is similar to the wavelength of visible light—color is generated. Multiple reflections from each layer or subdivision are responsible for this phenomenon, and the subsequent interference of reflected lights determines which colors are seen. See Chapter 1 for *interference*. The iridescent colors also change with the angle at which the objects are seen.

FIGURE 9.2 ▶ The Brazilian butterfly *Morpho peleides* has intense blue-cyan wings due to iridescence.

Iridescent microscopic structure is found on the surfaces of some animals, including bird feathers, beetle shells, and butterfly wings, as just mentioned. These animal structures are optimized to provoke destructive interference for all wavelengths of white light except green and/or blue. The latter wavelengths undergo constructive interference and are then visible to our eyes.

Let us look at the details of iridescence in the wings of the Brazilian butterfly. Light is reflected by arrays of microscopic ribs found on the wings of this butterfly. These ribs act as tiny mirrors, which reflect light. The size and spacing of the ribs ensure that only green and blue wavelengths interfere constructively and that red, orange and yellow wavelengths interfere destructively, canceling each other out. The same mechanism explains the color of the peacock, parakeet, kingfisher, and mallard feathers. All these feathers use small rods instead of ribs, but the mechanism is the same. The hummingbird feathers instead use oval-shaped platelets to reflect light. Thin-layered scales on the shells of green beetles also generate interference and iridescence.

If you were to alter or destroy the microscopic structure of these ribs, rods, platelets or scales—for example, by grinding the butterfly wings or bird feathers very finely—the color would disappear because there is no blue-green pigment present.

More Color-Generating Mechanisms Due to Iridescence

Iridescent colors can also be seen in soap bubbles, oil floating on water, pearls, opals, and other gems.

There are only two surfaces generating iridescence in a soap bubble, so this is the simplest of the iridescent systems. Light waves are reflected by both the outer and inner surfaces of the soap film that forms the bubble. The light wave reflected by the outer surface interferes with the one reflected by the inner surface. If these two waves are counterphase, as in Figure 9.3, they cancel each other out because of destructive interference. If they are in phase, however, they interfere constructively. Consider white light coming from a specific angle and a soap bubble whose film thickness is such that only green wavelengths cancel. In this case, both red and blue wavelengths are reflected and then added together, causing the surface of that portion of the soap bubble to appear magenta.

The thickness of the soap bubble is constantly changing; the top becomes thinner and thinner until the bubble bursts, and the water-soap mixture also moves because of convection. Furthermore, the surface of a bubble is curved, and therefore the film thickness through which light travels varies with the angle of incidence. All these factors combine to guarantee that all colors of the rainbow, plus the magentas and purples appear on a bubble's surface, as shown in Figure 9.4.

The same effect takes place on thin layers of oil floating on water, as shown in Figure 9.5. In Figures 9.6 and 9.7 we show the colors generated by iridescence in a biomineral and a mineral: mother-of-pearl and opals, respectively.

Light reflected by the nanostructures (identical parallel layers of aragonite in mother-of-pearl, or three-dimensional ordered arrays of identical silica nanospheres in opal) interferes destructively or constructively for different wavelengths, generating the colors seen on the shell and the opals in Figures 9.6 and 9.7. Interestingly, a rotation of the shell or the opal in front of the observer makes the same locations change color. This phenomenon is also called "opalescence" and is due to the change in effective spacing of the

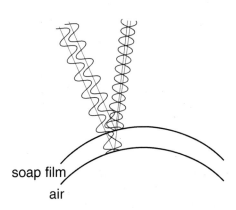

soap film
air

FIGURE 9.3 ▶ **Iridescence** in a soap bubble.

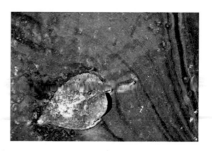

FIGURE 9.4 ▶ Part of a soap bubble with enhanced iridescent colors. Notice that magenta is *not* a spectral color. It results from the additive combination of red and blue light.

FIGURE 9.5 ▶ A thin layer of oil on water.

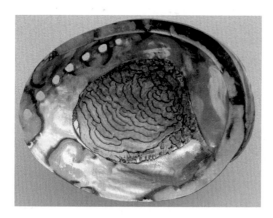

FIGURE 9.6 ▼
Mother-of-pearl (or nacre) in the inner side of a green abalone shell. Iridescence is generated by the regularly repeating, 400-nm-thick layers of aragonite ($CaCO_3$) tablets.

nanostructures when this is rotated. Pearls (Figure 9.8) are also made of mother-of-pearl like abalone, and their nanostructure is regularly layered so that each semitransparent aragonite layer is reflecting part of the light and transmitting the rest to lower layers.

Interference of the light reflected by subsequent layers generates the iridescence, which in pearls is called "pearlescence." Although Japanese pearls are white and Tahitian pearls are dark gray due to different composition, they both have extensive iridescence. Therefore, in both variations you can see a wealth of changing colors as you move them. Pearls are produced by a variety of organisms, most commonly sea mollusks (such as pearl oysters). They are also produced by freshwater mussels and, occasionally, by snails. Pearls are produced as equally-spaced layers of aragonite crystals, growing around an irritant such as a grain of sand, a piece of shell, or a parasite. Only the pearls

FIGURE 9.7 ▼
Australian boulder opals, formed by nanospheres of amorphous silicon oxide (SiO_2). These are arranged regularly into gratings, and spaced at distances between 400 and 700 nm (similar to the wavelength of visible light).

FIGURE 9.8 ▼
Tahitian pearls and Japanese white pearls.

from mollusks with a nacreous mother-of-pearl lining are "true" pearls. Most pearls on the market are cultured—that is, produced by inserting a marble bead into a pocket, inside the body of a pearl-producing mollusk. This is the case for Tahitian pearls. The mollusk treats the bead as an irritant and deposits a nacreous coating over it through a process that takes approximately two years. Chinese fresh-water pearls are instead produced by a mussel. The seeds (typically a piece of another "sacrificial" shell) are placed between the animal and the shell mother-of-pearl lining, and it takes ten years for the pearls to grow around the seeds. On the other hand, more than 40 pearls can be grown in a single mussel.

On rare occasions, iridescence can also be seen in clouds. When this happens, clouds are formed by myriad droplets of water, all having the same size of 1 micron (1000 nm). In these conditions the droplets act as in the rainbow (see Chapter 2), but the clouds display much more intense colors than rainbows against clear skies. Figure 9.9 shows stunningly colorful iridescent clouds.

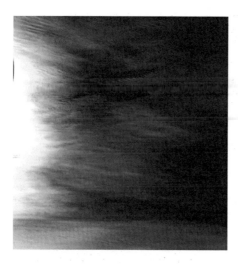

FIGURE 9.9 ▶ Iridescent clouds, reproduced from the website < *http: // www.psi.edu/iridescent.html* > with permission from David Lien.

The colors on a compact disc (CD), such as the one at the beginning of this chapter and on the cover of this book, are also structural colors, due to concentric circular lines of pits, with size and spacing similar to the wavelength of visible light. They therefore separate the colors of the spectrum at different angles.

In conclusion, we point out the different structures that generate iridescent colors in different systems. These include: *thin layers* in soap bubbles, oil spills, platelets in hummingbird feathers, scales on green beetles, and mother-of-pearl; *spheres* in opals; *rods* in peacock feathers; *ribs* in the morpho butterfly; and *pits* in CDs and DVDs.

9.5　Color in Gemstones

There are several mechanisms that generate color in minerals. Besides the obvious ones based on absorption of certain wavelengths and scattering of others that we have already discussed, color can come from the presence of single-atom impurities, their concentration, the specific sites in the crystalline structure of a mineral in which these impurities reside, the oxidation state of the impurities, and so on. For example, pure corundum is an oxide of aluminum (Al_2O_3). It is a colorless mineral. However, if corundum contains even a small amount of chromium (Cr), it becomes pink or red (and is then called ruby). In this case, the impurity is the presence of Cr. Chromium impurities in different minerals, however, cause ruby to be red, emerald to be green, and (most interestingly) alexandrite to assume different colors (purple, red, green, or blue), depending on the illuminating light!

Figure 9.10 shows all these gemstones, which are differently colored by the same element, chromium.

Different levels of impurity (1–2%) in the same mineral (e.g., beryl, which is $Be_3Al_2Si_6O_{18}$) generate different colors. Figure 9.11 shows all the different colors of beryl gemstones. Here are some examples of beryl gemstones, with different impurities.

FIGURE 9.10 ▶   Ruby (top left), emerald (top right), and alexandrite illuminated by fluorescent light (bottom left) and candlelight (bottom right). All colors in these different minerals are due to chromium impurities. Rough emerald is the author's photo. Ruby and Alexandrite photos are courtesy of Jillian F. Banfield and Hanna Cook-Wallace, from the course website *http://socrates.berkeley.edu/~eps2/*.

FIGURE 9.11 ▶   Beryl gemstones have different colors due to different metal atom impurities (Fe, Mn, or Cr) and the oxidation state of these metal atoms. Photos courtesy of Jillian F. Banfield and Hanna Cook-Wallace, from the course website *http://socrates. berkeley.edu/~eps2/*.

BERYL

Aquamarine = blue beryl: Fe^{2+}; this beryl is cyan
Heliodor = golden beryl: Fe^{3+}; this is yellow
Green beryl: mixtures of Fe^{2+} and Fe^{3+} are green
Morganite: Mn^{2+} is pink
Red beryl: Mn^{3+} is red
Emerald: Cr^{3+} is . . . emerald green!

The alexandrite effect (a color change as the result of a different type of illuminating light, as shown in Figure 9.10) is an excellent example of how transmission and absorption of specific wavelengths determine color. Light from a lightbulb on a dimmer or a lit candle is rich in red and yellow wavelengths while fluorescent light is rich in blue and green. Alexandrite can transmit and reflect blue, green, and red light. Therefore, if the light illuminating alexandrite is rich in red and blue, it will transmit both red and blue and appear purple. On the other hand, if the illuminating light is rich in red but poor in blue and green (candlelight), this same piece of alexandrite will look red. If the light is rich in blue and green but poor in red (fluorescent lights), it will look blue or green.

9.6 Mineral Color Due to Charge Transfer

Another mechanism generating color in gemstones is charge transfer. Charge transfer involves the exchange of an electron between two atoms. For example, blue sapphire (Figure 9.12) is colored by iron (Fe) and titanium (Ti), and it appears blue because R and G light wavelengths are absorbed to transfer electrons back and forth between Fe and Ti atoms.

9.7 Mineral Color Due to Color Centers

Color centers are very small (atomic-scale) defects in the crystal structure of the gemstone. Even a single electron, when trapped within the crystal structure of a gemstone, is a color center. The electron absorbs some wavelengths of light and uses this energy to "jiggle" within its trap. The remaining wavelengths are scattered, generating the color of the gemstone.

Since these color centers are defects of the crystal structure, they can be "healed" by heating the gemstone or exposing it to UV light. Heat allows the crystal to reorganize itself, while UV light frees the electrons trapped in color centers. Both processes result in removal of the color center, causing the gemstone to change color. If the only source of color was the color center, the stone will become colorless after heating or exposure to UV light.

FIGURE 9.12 ▼
Blue sapphire, which is corundum (Al_2O_3) with Fe^{2+} and Ti^{4+} atoms that exchange electrons to become Fe^{3+} and Ti^{3+}, and vice versa. Photo courtesy of Jillian F. Banfield and Hanna Cook-Wallace, from the course website *http://socrates.berkeley. edu/~eps2/.*

FIGURE 9.13 ▼
Swiss blue topaz, obtained by irradiation with either neutrons or high-energy electrons and subsequent heat treatment of natural white topaz.

Exposure of gemstones to high-energy particles produces color centers. High-energy particles include the following:

- Accelerated electrons
- Accelerated alpha particles
- Gamma ray photons from a radioactive cobalt-60 source
- Neutrons from a nuclear reactor.

Color centers give color to naturally yellow topaz. White topaz can also be irradiated and heat-treated to make it blue (London blue, Swiss blue, or sky blue), brown, or green. Figure 9.13 shows Swiss blue topaz. After irradiation the gemstones are radioactive and cannot be used until the radioactivity has decayed to safe levels. This may take months or years, depending on the country of origin of the topaz, the irradiation energy, and the exposure time.

9.8   Color in Gems Due to Band Gap Absorption of Light

In materials such as semiconductors and metals, some electrons are no longer associated with individual atoms but are free to roam around the entire crystal. Just like in pigments and in color centers, electrons in these materials can absorb light at certain wavelengths, thereby generating color in the light they scatter. In semiconductors, however, there is an additional phenomenon: Light is only absorbed when its wavelength is below a threshold value. Recall that smaller wavelengths correspond to greater energy (see Figure 1.6), and for light to be absorbed by a semiconductor it must have enough energy to make the electrons jump over a forbidden energy gap, called the band gap. Light with too large a wavelength, therefore, is transmitted, and the material appears transparent. Materials that have a band gap at wavelengths greater than the visible range (in the infrared) appear completely opaque to us (e.g., silicon); those that have the band gap below the visible range (in the ultraviolet) are transparent (e.g., diamond).

If the band gap falls within the visible range, some colors will be absorbed and others transmitted and/or scattered, therefore the material appears colored. The color in this case may be yellow, orange, or red, depending on where the threshold wavelength lies in the optical spectrum, but not blue. Because the transmittance and reflectance are never large at low wavelengths and low at high wavelengths, blue is impossible in pure semiconductors.

The presence of impurities, however, may introduce additional electron energy states within the forbidden band gap. For instance, impurities in diamonds provoke band gap states that absorb specific wavelengths and generate color. In diamonds, the presence of nitrogen impurities generates a yellow color, while boron generates a blue color. Figures 9.14 and 9.15 show diamonds with a variety of colors.

FIGURE 9.14 ▶ Diamonds of multiple colors. Yellow and brown diamonds contain nitrogen impurities in the band gap. Gemological Institute of America. Reprinted by permission.

FIGURE 9.15 ▶   The famous Hope Diamond, valued at more than $200 million, is a large (45.5 carats) blue diamond, made blue by boron impurities in the band gap. It is surrounded by other colorless diamonds. Reproduced with permission from the National Museum of Natural History, The National Gem Collection. Photo by: Dane A. Penland. Copyright © 2007, Smithsonian Institution.

Chapter 10

Periodic Oscillations

Why do we hear something when we tap a tuning fork (see Figure 10.1) with a rubber hammer? Let's start with a rough explanation of the sequence of events:

1. The two ends of the fork (tines) swing toward and away from each other rapidly. The motion is too small and too rapid to be followed by eye, but by touching the tuning fork lightly, you can feel the vibrations. For the "standard A" tuning fork, used by the concertmaster for tuning the orchestra, there are 440 to and fro oscillations every second.

2. As the tines of the tuning fork swing back and forth, air near the tuning fork is pushed in the corresponding direction. This causes a slight variation in air pressure near the tuning fork that is in step with the motion of the fork.

3. The pressure variations spread out in the surrounding air, much like the ripples on the surface of a lake spread out when you toss a rock into the water.

FIGURE 10.1 ▶ A tuning fork.

Physics in the Arts, Revised Edition

4. When the pressure variations reach your ear, the eardrum is moved, ever so slightly, in step with the pressure variations.

5. The motion of the eardrum is converted to nerve pulses, which your brain interprets as the sound of a tuning fork. Your brain does not register each oscillation individually, but the oscillations are encoded in the ear in an interesting way, which we will discuss in Chapter 16.

In order to understand sound—how it is produced, transmitted through the air, and perceived—one must begin by understanding mechanical oscillatory motion. Let us first learn what makes objects *oscillate*.

Vibrations, or oscillatory motions, are not unique to sound. Oscillations are quite common in nature. Familiar examples are slow mechanical oscillations: a swinging pendulum, a weight bouncing up and down at the end of a spring, a squirrel swinging up and down after jumping on a tree branch, or a plucked guitar string. In all oscillations there is a to and fro motion. What else is common to all oscillations?

1. *A position of equilibrium*: A place where the oscillating object remains at rest if it is left alone. In the case of the pendulum, the equilibrium position is at the bottom of its swing.

2. *A restoring force*: A force that always pulls the object back to its point of equilibrium once it is displaced from its equilibrium position. If you pull a person on a swing away from their point of equilibrium, gravity will pull them back.

3. *Inertia*: When a mass is in motion, it keeps going (think of a spacecraft in outer space) unless there is a push or pull to change the velocity of motion. We say that the mass has "inertia," i.e. the mass resists a change in its state of motion. Therefore, if an object is stationary, it will stay put. If it is moving, it will keep moving until another force stops it. Without inertia, a person on a swing would not keep moving back and forth but simply return to the point of equilibrium and go no further. (Swings sure wouldn't be much fun!)

As a simple example, let us look at a time sequence of the motion of an oscillating object, such as the one in Figure 10.2.

A more quantitative statement about motion is expressed by Newton's law:

$$\boxed{\vec{F} = M \cdot \vec{a}}$$

The equation in words is *force equals mass times acceleration*. Acceleration *a* means how much the velocity changes per second, and force means

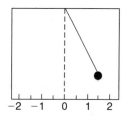

$t = 0 \sec, \; x = +1.5 \, \text{m}$
Somebody pulled the pendulum to the right and released it. If we count the displacement to the right of the equilibrium position as positive, the initial displacement shown in the picture on the left is $x = +1.5 \, \text{m}$.

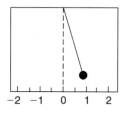

$t = 1 \sec, \; x = +1 \, \text{m}$
Gravity applies a downward force on the mass, but the string keeps the mass from falling straight down. Instead, the mass is pulled back toward the lowest point (equilibrium position or rest position). The pull of gravity speeds up the mass. We say that the *force* acting on the mass *accelerates* the mass.

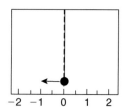

$t = 2 \sec, \; x = 0$
The pendulum is at the equilibrium position but keeps moving because of inertia. There is no braking force to stop it.

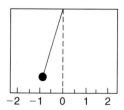

$t = 3 \sec, \; x = -1 \, \text{m}$
Now the pendulum is being slowed down by the force of gravity, which pulls the pendulum to the *right*, toward the equilibrium position, but the pendulum still keeps going until . . .

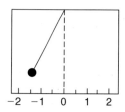

$t = 4 \sec, \; x = -1.5 \, \text{m}$
the continuous downward pull of gravity stops the pendulum at $-1.5 \, \text{m}$. But gravity again pulls it back toward the equilibrium position, and so on. You can complete the sequence on your own.

FIGURE 10.2 ▶   The position x of a pendulum at different times t.

the amount of push or pull on the mass M. Because acceleration means the amount of velocity change in *one* second, after t seconds the change in velocity would be

$$\boxed{\Delta \vec{v} = \vec{a} \cdot t}$$

change in velocity = acceleration times how many seconds (t) the force acted.

The Greek symbol Δ (delta) is commonly used to denote a *change* in a quantity. The arrow over the symbols means that the quantity has a particular direction—that is, if the force is in the downward direction, then the change in velocity is also in the downward direction. Note that the velocity might well be upward at a given instance in time (e.g., a ball thrown vertically upward), but gravity (a downward force on the ball) would tend to pull the ball downward, and the velocity will therefore decrease. The upward velocity decreases every second until the ball stops in midair and starts falling back down.

The acceleration of gravity on Earth is 9.8 m/s^2. This means that for an object in mid-air after, for example, 2 seconds, the pull of gravity adds an additional velocity of $(2 \text{ s}) \times (9.8 \text{ m/s}^2) = 19.6$ m/s in the downward direction.

Exercise: Use Newton's law to find the force of gravity on a 5-kilogram bag of apples.

Answer: Mass = 5 kg, a = acceleration of gravity = 9.8 m/s^2. F = 5 × 9.8 kg m/s^2 = 49 kg m/s^2

For brevity, 1 kg m/s^2 is called 1N (pronounced "one Newton"). So the answer is **49 N**.

Example: If you throw a ball up vertically with a velocity of 30 m/s, what will the vertical velocity be after two seconds? Well, you had 30 m/s upward to start with, and you added $(2 \text{ s}) \times (9.8 \text{ m/s}^2) = 19.6$ m/s in the downward direction, so there will be $(30 - 19.6) = 10.4$ m/s left in the upward direction.

10.1   Displacement Graph: Position *x* Changes with Time *t*

Let us make a graph showing the position x of the pendulum in Figure 10.2 at various times t. We will plot the *displacement x*—that is, the distance from equilibrium position (in meters) versus *time t* in seconds (s). We choose to call displacement to the right of the equilibrium position *positive* and to the left *negative*.

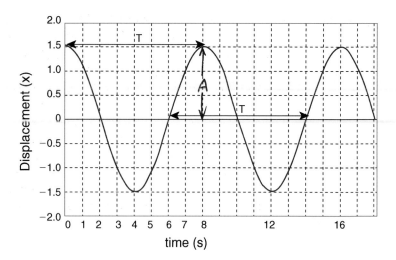

FIGURE 10.3 ▶   The graph shows how the displacement x changes with time. Here we plotted time on the horizontal axis of the graph and displacement x vertically.

We are going to simplify matters, as shown in Figure 10.3, and only think about the motion of the pendulum in the x direction (parallel to the ground). Note that a graph is an abstraction: The actual motion of the object may be horizontal, but that does not keep us from plotting the displacement on the vertical axis.

Looking at the graph in Figure 10.3, one can make several observations:

1. The pendulum swings at most 1.5 m to either side of the rest position. This distance is called the ***amplitude A*** of the oscillation, just like in Figure 1.4 for light waves. Here $A = 1.5$ m.

2. The motion we are looking at here is called ***periodic***, because it repeats itself over and over again. The time taken for one complete back-and-forth oscillation is called the ***period T***. We see from Figure 10.3 that the period in this example is $T = 8$ s, where s (or sec) is used as an abbreviation for "seconds." Note that it does not matter where one starts counting the time: T is marked in two different but completely equivalent ways in the graph.

3. Instead of referring to the period T, one often speaks of the number of oscillations (or cycles) per second. The number of complete back-and-forth oscillations per second is called the ***frequency f***. The unit of frequency is called ***Hertz***, abbreviated ***Hz*** (named after a German physicist who studied high-frequency oscillations, which were later used for radio transmission). So instead of saying that a tuning fork has a frequency of 440 oscillations per second, we say that the frequency is 440 Hz

(pronounced "440 hertz"). Years ago, the abbreviation "cps" was used for "cycles per second" but the current international convention is to use Hz.

4. *Different wave shapes:* If we plot the motion of a tuning fork, we see a graph very much like Figure 10.3, except that the time scale is much faster. The simple wave shape shown in Figure 10.3 is referred to as a *sinusoidal motion* or as *simple harmonic* motion. (We'll discuss what that means in the next chapter.) The tone produced by simple harmonic motion of the tuning fork in music is called a **pure tone**. Most musical instruments (and the voice) make more complicated oscillations when they play a steady tone. The oscillations are periodic, but the shape of the pressure variation versus time is different for each instrument. Two examples of *periodic* but *non-sinusoidal* oscillations are shown in Figure 10.4.

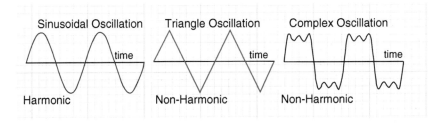

FIGURE 10.4 ▶   Graphs representing a harmonic and two non-harmonic oscillations.

10.2   The Period *T* and the Frequency *f*

As you may have imagined already, the period and frequency are related quantities. If the period is very short, there are lots of oscillations in one second and thus the frequency is high.

  If there are *f* oscillations in one second, then the time *T* required for each oscillation must be $1/f$ seconds. Mathematically:

$$\boxed{T = 1/f} \quad or \quad \boxed{f = 1/T}$$

Anything that oscillates with a *high* frequency *f* must have a *short* period *T*, and vice versa.

Example: The "standard A" tuning fork that is used in tuning instruments has a frequency of 440 Hz. Find the period of this tuning fork.

Answer: $T = \frac{1}{440\,\text{Hz}} = \frac{1}{440(1/\text{s})} = \frac{1\,\text{s}}{440} = 0.00227\,\text{s}$

10.3   Large and Small Numbers

In the previous example, you already noticed that the period turns out to be a small number with several zeroes after the decimal point. The numbers get even smaller when the frequency is high, such as a dog whistle, which may have a frequency of 30,000 Hz.

It is convenient to express large and small numbers by the use of prefixes. Some of these you already encountered in the chapters on light, such as nanometers (nm) when you specified wave lengths.

	Prefix	Such as in	Example
$1000 = kilo\text{-}$	k	kg (kilogram), kHz (kilohertz)	30,000 Hz = 30 kHz
$1,000,000 = mega\text{-}$	M	MHz (megahertz)— FM station	88,700,000 Hz = 88.7 MHz
$10^9 = giga\text{-}$	G	GHz (cordless phone)	2.7 GHz = 2,700,000,000 Hz
$1/100 = centi\text{-}$	c	cm (centimeter)	Person height 1.7 m = 170 cm
$1/1000 = milli\text{-}$	m	ms (millisec), mg (milligram)	0.00227 s = 2.27 ms 0.03 g medication = 30 mg
$1/1,000,000 =$ $10^{-6} = micro\text{-}$	μ	μs (microsecond) μm (micrometer)	Diameter of human hair 0.000,05 m = 50 μm
$1/1,000,000,000 =$ $10^{-9} = nano\text{-}$	n	nm (nanometer)	Diameter of a gold atom 0.000,000,000,3 m = 0.3 nm

10.4   Speed of Motion

By looking at the graph (Figure 10.3), one can deduce how the object speeds up and slows down. For instance at $t = 1.5$ s, the displacement x is about +0.5 m, and a half second later ($t = 2$ s) the position is $x = 0$. Thus, the average speed during this time interval is $v =$ (distance covered)/(time used) $= 0.5$ m/0.5 sec $= 1.0$ m/sec. Here we don't care about the direction of motion; we simply ask how far the object moved in 1s. We are talking about the *speed* of the object, while *velocity* also specifies the direction of motion. If we count velocity as positive in the $+x$ direction, it would then be negative in the $-x$ direction. If a car goes into a curve at constant speed (using cruise control), does the speed change? No, it doesn't. Does the velocity change? Yes, because the direction of the motion is changing.

What we have used here is the concept of speed as you know it from driving a car: the speedometer tells you the distance covered per hour if you were to keep going at the same rate for the entire hour. What the speedometer reads is what we call the instantaneous speed (the speed at that instant). Even though it is labeled as miles per hour (mph), it does not imply that we keep going for an entire hour. Mathematically we express the speed as

$$speed = \frac{\text{distance}}{\text{time}} = \frac{\Delta x}{\Delta t}$$

where Δt means a small time interval, and Δx is the distance covered during this time interval.

You might well say that science is not supposed to use wishy-washy concepts like "small time interval" because how would we decide if a time interval is "small"? And "small" compared to what? Well, the idea is that the time interval is small enough that the motion can be considered steady during this time interval. So the required smallness depends on whether the motion is very smooth, changing slowly in time, or very abrupt and changing rapidly. Look at the displacement graph in Figure 10.5: near the maxima and minima of the curve, the displacement graph can be considered nearly straight only for a very short interval of time, but where the displacement passes through zero a larger interval is permissible.

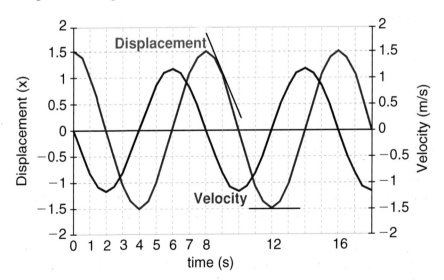

FIGURE 10.5 ▶ The magenta curve, which starts out at $+1.5$, shows the displacement—that is, how the position changes with time. The blue curve shows how the velocity changes with time. The velocity scale is shown on the right vertical axis of the graph. Note that the velocity is largest when the pendulum passes through its rest position ($x = 0$).

A nicer way to visualize the meaning of the speed at one particular instant is to draw a tangent to the graph at that instant. In Figure 10.5 we show the tangent at $t = 9$ s and at $t = 12$ s. At $t = 9$ s the tangent points downward (negative velocity) and has a slope of -0.8 m/s, while at $t = 12$ s the tangent is horizontal, which indicates that at that time, the mass is at rest for an instant.

If you think of the slope of the curve in Figure 10.5, you find that the curve is steepest at $t = 2$ s, $t = 6$ s, and $t = 10$ s — in other words, it is steepest when the displacement x is zero. This should make sense to you if you remember that the graph in Figure 10.5 represents the motion of a pendulum. Imagine a swinging pendulum and ask yourself at what position does it move most rapidly. Well, it is fastest at the bottom of the swing when it passes though the equilibrium position!

Some of you may have studied calculus. In calculus you learned that as you make the time interval Δt shorter and shorter (and thus Δx also gets smaller and smaller), eventually the ratio $v = \Delta x / \Delta t$ reaches a limit. Strictly speaking, it is this limit that is called the instantaneous speed: $v = dx/dt$—that is, the **instantaneous speed is the derivative of x(t).** This exactly corresponds to the tangent line we just mentioned. The displacement graph we showed in Figure 10.5 (magenta curve) is a cosine graph; if you take the derivative of the cosine, you get a negative sine graph, just as we showed in Figure 10.5.

10.5   Questions

1. The period of an oscillation is 4 ms. What is its frequency in Hz and in kHz? If the frequency of oscillation is 1.3 MHz, what is the period in ms and in μs?

2. A dog can hear frequencies up to 40,000 Hz. Express this frequency in kHz. Find the period of this oscillation (express the answer in s, in ms, and in μs).

3. For the oscillation shown here, the horizontal axis (time) is in *msec*. Find the period T in *sec* and in *msec*. Find the frequency in Hz and in kHz. Find the amplitude.

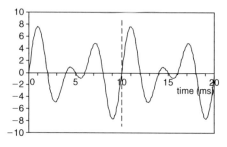

The graph shows the sum (superposition) of a 200 and 300 Hz tone.

4. Conversion of units:

   2 km = _____ m   1.5 cm = _____ m

   25 mm = _____ cm   2600 g = _____ kg

   7800 Hz = _____ kHz   0.0003 s = _____ ms

5. The range of a young person's hearing extends from 20 Hz to 20 kHz. Find the corresponding periods of oscillation.

6. A violinist plays a very fast passage in which each tone lasts only 0.2 s. How many to and fro oscillations does the string make during this time when the frequency of the tone is (a) 200 Hz and (b) 1000 Hz?

► Chapter 11

Simple Harmonic Motion

Oscillatory motion requires something elastic or "springy." Deformation of the object causes a restoring force, and this force drives the object back toward the original shape—in other words, toward the equilibrium position. Think of a tree branch: If you pull down on the branch, it yields a certain amount—there is a "displacement." The harder you pull, the larger the displacement. A thick, heavy branch will yield less, and a thin branch will yield more. It is the same with guitar strings: When you pluck the string, you displace it from the equilibrium position. When you release the string, it snaps back—there is a restoring force.

How do we measure the restoring force, and how does the restoring force affect the oscillation when you release the branch or the guitar string?

To keep things simple, we use as a springy object a spring or—if you prefer—a rubber band. To find out *how* springy the spring is, we hang a weight from it and see how much it stretches (see Figure 11.1). A "soft" spring would yield a lot, while a "stiff" spring stretches only a small amount.

How do we measure the stretching (called the ***elongation*** x)? Simple! We use a ruler and measure how much longer the spring became as we added the weight. Suppose the spring or rubber band was 30 cm long to start with, and after the weight was added, it was 35 cm long. The elongation in this case would be $(35 - 30)\,\text{cm} = 5\,\text{cm}$. Usually we measure the length in meters (m), rather than centi-meters (cm). Since *centi-* means 1/100, $5\,\text{cm} = (5/100)\,\text{m} = 0.05\,\text{m}$.

How do you measure the downward pull (***weight***) of the mass attached to the spring? If you attach a mass to the spring, why does the spring get longer?

Physics in the Arts, Revised Edition

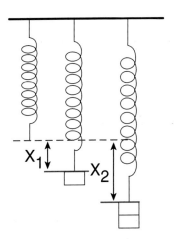

FIGURE 11.1 ▶ Different masses are attached to a spring. If each mass is 1 kg, the force of gravity (weight) on each mass is 9.8 N.

You will say that Earth pulls down on the mass, and this downward pull is what we call the *weight* of the mass. Why be so fussy and distinguish between **mass** (how big a chunk you attached to the spring) and **weight** (how much the mass gets pulled down by gravity)? Well, the mass is simply the amount of matter, but the weight depends on where you are. On the moon, the same mass gets pulled down much less than on Earth because the moon is so much smaller, while in outer space, away from all planets, the same mass would be weightless. A conceptually simple way to measure the weight of a mass, be it on Earth or on the moon, is to let go of it. The mass will pick up speed during the fall (it will accelerate), but on the moon it will pick up much less speed every second because the pull is less. If you measured the speed after 1 s, 2 s, 3 s, you could find the *acceleration of gravity* (called g), meaning how much additional speed the mass picks up every second. As we mentioned in the preceding chapter, on Earth, $g = 9.8 \, m/s^2$, meaning that every second the speed of a falling object increases by that amount. The units come about from acceleration being a change in speed (m/s) divided by the time (s), so it makes m/s^2.

To measure the amount of pull (weight), we use Newton's law of motion, $F = M \times a$.

If we can measure the mass and the acceleration, we know the pull of gravity, F. Here, the acceleration of gravity is called g instead of a, so the weight W (the force F of gravity on the mass M) is

$$\boxed{\text{Weight of mass } M \text{ on Earth: } F = M \cdot g}$$

Since it may be hard to keep track of all of this, let us simply state the end result:

When you have to figure out the pull of gravity (i.e. the weight) on a mass M, you simply multiply the mass (in kg) by 9.8 and you get the weight in N (Newtons).

Example: If Willy's mass is 75 kg, how large is his weight (on Earth)?

Answer: His weight is $75 \times 9.8\,\text{N} = 735\,\text{N}$.

Example: If Willy jumps off a cliff, how large is his (downward) acceleration?

Answer: According to Newton's law, $F = Ma$. Here, $F = 735\,N$, $M = 75\,\text{kg}$, so his downward acceleration is $a = F/M = 735/75\,\text{m/s}^2 = 9.8\,\text{m/s}^2$. Of course, the answer must be just the acceleration of gravity. The acceleration is the same for any mass, be it Willy's or any other object or body jumping off a cliff!

11.1   The Spring Constant

To test springs, we hang different weights from the spring, and for each weight we record how much the spring got stretched compared to its original length. Suppose we increase the weight step by step by adding an additional 0.5 kg mass each time. The results might look something like numbers in the following chart.

mass M (kg)	0	0.5	1.0	1.5	2.0
weight (N)	0	4.9	9.8	14.7	19.6
elongation x (cm)	0	1.5	3.0	4.5	6.0

The thing to note is that equal increments in mass (i.e., equal increments in weight) make equal increments in the length of the spring. Or *the elongation of a spring is proportional to the pull*. The graph of spring length versus force pulling on the spring is a straight line. Of course, this is true only up to a point: if you pull too hard on the spring, eventually it will be stretched out of shape.

In fact, there is usually some small measurement error so that the numbers may not be as perfect as in this table.

If F is proportional to elongation x, the results can be expressed as

$$F = k \cdot x \qquad \textbf{or} \qquad k = F/x$$

The quantity **k** is called the *spring constant*.

In the preceding example, for $F = 4.9N$ pull on the spring, the elongation was 1.5 cm. The spring constant in this case is $k = 4.9$ N/1.5 cm = 3.27 N/cm.

If we convert the elongation to *meters* instead of cm, the answer is $k = 4.9$N/(0.015 m) = 327 N/m.

All other columns of the table would give the same result because F and x increase in proportion.

Not all springy things have a spring *constant;* in other words, not everything stretches evenly in a straight-line fashion when you increase the pull. If you use a rubber band instead of a metal spring, you will usually find that with equal increments in pull, the rubber band yields *more* than it should because the rubber band gets thinner and thinner, making it weaker and weaker, thus easier to stretch.

Stiff springs have a large value of the spring constant because a stiff spring requires a larger force than a soft spring for a given amount of stretching. Until now, we talked about stretching or elongation of a spring, but the same applies to compression of a spring, such as the springs of a car suspension.

11.2   Oscillation Frequency for Simple Harmonic Motion (SHM)

Simple Harmonic Motion is the name given to the motion that results when the force on a mass is exactly proportional to the displacement of the mass from its rest position.

When the mass hanging from a spring is displaced and then released, it will oscillate. How rapid is this oscillation, what does the oscillation frequency depend on, and what is the wave shape of this oscillation? The answers to these questions will be important later on when we want to figure out what musical tone is produced by the strings of a piano or a guitar.

You are perfectly correct if you guessed that, with some math, one can figure out the frequency of oscillation of a mass on a spring, given that we know the *spring constant k* and the amount of *mass M*. Newton's law, $F = M \times a$, says that if we know the spring force F and the mass M, we can figure out the acceleration. Knowing the acceleration, in turn, lets us calculate the velocity at each instant in time, so we can predict how the object moves, based on a simple calculation. But what makes the calculation complicated is that as the mass moves up and down, the position changes, which changes the spring force, which changes the acceleration, which changes the velocity, which changes the position—and so we are back to the beginning of the chain of quantities, all of which change with time.

The math required for this calculation is rather elementary calculus. But instead of just telling you the result, let's see if we can figure out at least part of it together. Whenever possible, you should try to reason things out rather than just taking somebody's word for it.

When you displace the mass some distance x, there will be a restoring force F pulling the mass back to the equilibrium position. A large spring constant k means the force is large, causing a large acceleration. The mass picks up speed quickly and returns to the equilibrium position in a short time. Conclusion: *Large spring constant k causes rapid oscillations (large frequency).* On the other hand, a large mass is sluggish, picks up speed slowly, so for the same spring as before, *a larger mass will produce a lower frequency of oscillation*. These two statements imply that in the formula for frequency, the spring constant k should be in the numerator and the mass M should be in the denominator—larger k gives larger f, larger M gives smaller f.

So we would think that the vibration frequency depends on k/M. Basically, this is correct, but let us look at the units: The frequency must have units of number of oscillations per second, but does this agree with our answer k/M? If you figure it out (k has units of N/m, the unit N, for Newton, equals kg × m/s^2, M has units of kg) you find that our answer has units of $1/s^2$, not $1/s$. So we need to take the square root of the quantity k/M.

Indeed, the correct answer for the oscillation frequency of a harmonic oscillator is

$$f = \tfrac{1}{2\pi}\sqrt{\tfrac{k}{M}}$$

This is called the **natural frequency** of the oscillator, *natural* in the sense that if the mass M is set in motion and then left to itself without outside interference, f is the frequency at which the mass inherently wants to oscillate. The formula is just what we guessed; the only thing we missed in our guess is the $1/2\pi$ in front. For the formula to work out, we must use the proper units: k is expressed in **N/m** and M in **kg**. The frequency then has units of $1/s$, which means so many complete oscillations per second, or so many Hertz.

Example: At a farmers' market, a farmer uses a spring scale to weigh apples. Assume that 2 kg of apples stretch the spring by 10 cm. Find the oscillation frequency of the apples on the spring scale after they are given a push.

Answer: The weight is $2 \times 9.8\,\text{N} = 19.6\,\text{N}$. The elongation is $10\,\text{cm} = 0.1\,\text{m}$, thus the spring constant is $k = F/x = 19.6\,\text{N}/0.1\,\text{m} = 196\,\text{N/m}$.

The frequency is $f = \tfrac{1}{2\pi}\sqrt{\tfrac{196}{2}}\,Hz = 1.58\,\text{Hz}$

Does amplitude affect the frequency? Our first intuitive reaction might be to say that larger amplitude means that the mass has further to travel on each swing so that the period (time for one oscillation) should get longer. Not so! If you make the initial displacement—for instance, twice as large—indeed the mass has twice as far to travel, but at the same time, for the larger displacement the spring force is also twice as large. Twice a spring force means twice the acceleration, twice the speed picked up, which exactly offsets the larger distance traveled. So, indeed, this is the reason that in the formula for the vibration frequency, the amplitude does *not* appear! In other words, the **natural frequency does not depend on the amplitude of oscillation**.

Note, however, that frequency does not change with amplitude *only* for a simple spring force, which is exactly proportional to displacement. In all other cases the frequency changes when the amplitude changes. *It is a special property of simple harmonic motion that the frequency of oscillation is independent of amplitude.*

This is of enormous importance in musical instruments. In music, **amplitude** is related to **loudness** of the tone, while *frequency* of oscillation is the **pitch** of the tone. When we say that the frequency is "independent" of amplitude, we are saying that the pitch does not change when you play softer or louder. The amplitude of a piano string can be large or small, depending on how loud a tone is being played, but the frequency is still the same. If not, it would be a disaster, since the musical pitch would change every time the musician goes from soft to loud.

11.3   Wave Shape of Simple Harmonic Motion

To produce a graph (position vs. time) for simple harmonic motion (SHM), we could imagine attaching a pen to the oscillating mass and sliding a strip of paper at a constant speed under the pen. The result is shown in Figure 11.2.

What exactly is the shape of the graph that characterizes simple harmonic motion? Those of you who had some trigonometry might say, "Just call it a sine curve." Even if you had no trigonometry, in a couple of minutes you can learn that the sine-function is just a name for something very simple, as shown in Figure 11.3. Draw a circle with a radius 1 unit of length, and draw a radial line that is rotated counterclockwise by an angle α (alpha) with respect to the horizontal. The length of the green line in Figure 11.3 is called the "sine" of angle α, or $\sin\alpha$ (read "sine alpha"). No math skills are needed—just a protractor to measure angles and a ruler! In Figure 11.3, the angle is 30°, and you see that $\sin 30° = 0.5$. If instead you draw the radius at 45° and count

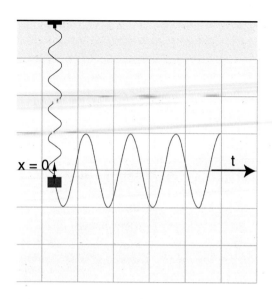

FIGURE 11.2 ▼
Wave shape of simple harmonic motion. A mass hanging from a spring oscillates and makes a trace on a paper strip that moves to the right. The horizontal axis shows time, and the vertical axis is displacement. It is fun to look at the *demonstration* on the Web: *http://surendranath. org/Apps.html.* In the "oscillations" menu select "forced oscillations" and set force amplitude and resistance to zero.

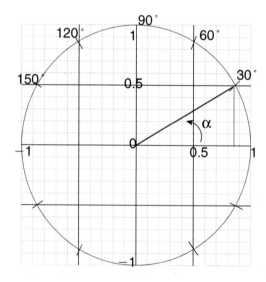

FIGURE 11.3 ▼
Explanation of the sin function. The circle has a radius $R = 1$. The red line is a radial line rotated counterclockwise by an angle α with respect to the horizontal. The length of the vertical green line is called "the sine of the angle α." In this example, you find $\sin 30° = 0.50$. You also note that $\sin 0° = 0$, and $\sin 90° = 1$. When the angle α is between 180° and 360°, the sine of the angle is negative.

the number of squares, you find $\sin 45° = 0.707$. You also see that the largest value of $\sin \alpha$ is found when $\alpha = 90°$; in other words $\sin 90° = 1$.

You might say, "Why make a graph like Figure 11.3? Didn't it occur to anybody to write a book with values of sine for different angles?" Yes, but even more convenient, most pocket calculators have a "sin" button. If you put in 45 and press "sin," it should show 0.707 on the display. If not, your calculator may use a different way to measure angles. In radians, once around a circle is 2π radians. In this case you have to reset the calculator to use degrees, so once around a circle is $360°$.

Exercise 11.1 Draw the radius line for $\alpha = 210°$ in the graph and show that $\sin 210° = -0.5$. It helps to note that $210° = 180° + 30°$

An equation to describe SHM: The displacement x of a pendulum (or the air pressure for a tuning fork making SHM) can be written as an equation that contains the amplitude A of the oscillating object, the period T, and the time t at which you want to know the displacement. The answer is

$$\boxed{x = A \cdot \sin(360° \, t/T)} \quad \text{or, if angles are in radians,} \quad \boxed{x = A \cdot \sin(2\pi \, t/T)}$$

Let's see if this makes sense: When one period has passed, the time t equals the period T, and thus $t/T = 1$. Indeed, after one period, the angle in the sin-function must be $360°$ (or 2π, if you use radians instead of degrees). The reason to multiply sin by the amplitude A is that the largest value of sin is 1, but the largest displacement is A.

Often we would rather express x in terms of the frequency f instead of the period T. We can simply replace $1/T$ by f:

$$\boxed{x = A \cdot \sin(360° f \cdot t)} \quad \text{or, if angles are in radians,} \quad \boxed{x = A \cdot \sin(2\pi f \cdot t)}$$

Exercise 11.2 A tuning fork vibrates with a frequency of 200 Hz and amplitude 1 mm. Make a graph showing the displacement every 0.5 ms between time $t = 0$ and time $t = 5$ ms. Use a calculator to look up the sin values.

Answer: It is a bit tedious but easier than you think! Make a table of time values t (0 s; 0.5 ms = 0.0005 s; 1.0 ms = 0.001 s; etc.). For each value calculate frequency times time. Then get sin of that number from the calculator, and multiply by A; this is the displacement x. The graph should look similar to Figure 11.2.

The displacement graph you calculated in this way is called a *sinusoidal oscillation* (see Figure 11.4).

11.4 Phase Angle

The angle that the rotating arm makes with the horizontal is called the phase angle. When we look at a displacement versus time graph for a harmonic

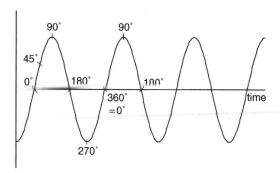

FIGURE 11.4 ▶ A sinusoidal oscillation. The graph shows displacement vs. time. The phase angle is marked at different points of the oscillation; it is the angle between the horizontal axis and the radial line in Figure 11.3. After a complete oscillation, the phase angle is 360° or 0°, whichever you prefer to call it.

oscillator (a sine curve), we can mark the phase on the curve or on the time axis. The phase angle is particularly useful later when we compare two similar sinusoidal oscillations that are out of step.

11.5   Questions

1. A spring is 7 cm long when no force is applied to it. When a 2 N pull is applied to the spring, it stretches to a total length of 13 cm.
   What is the spring constant for this spring?
   If a 0.5 kg mass is attached to this spring, what will the frequency of oscillation be?

2. A certain spring has a frequency of 100 Hz when a 50 g mass is hung from it.
   What would the frequency be if the mass were changed to 200 g?
   What would the frequency be if the mass were changed to 500 g?
   Does a *stiff* spring have a larger or smaller spring constant k than a soft spring?
   Does a *stiff* spring cause a higher or lower oscillation frequency than a soft spring?

3. If Jim has a mass of 70 kg, how many Newtons does he weigh on Earth where the acceleration of gravity is $g = 9.8$ m/s^2? If Jim is on the moon where the acceleration of gravity is only 1.62 m/s^2, how much will he weigh?

4. The length of a spring is measured for different pulls F:

   pull F: 2 N   4 N   6 N   8 N   10 N
   length of spring: 20 cm 25 cm 30 cm 35 cm 40 cm

   Find the spring constant in N/cm and in N/m.

5. A 2 kg mass oscillating up and down on a spring makes 10 oscillations in 5 sec, which corresponds to a frequency of ___ Hz. What would the frequency be if the 2 kg mass was replaced by a 4 kg mass?

► Chapter 12

Marimba from the highlands of Guatemala

Damped Oscillations and Resonance

12.1 Damped Oscillations—The Concept of "Damping Time"

Up to now we have talked about oscillations (pendulum, mass, and spring, etc.) as if—once put in motion—these oscillations continued indefinitely. We know that in the real world, oscillations sooner or later come to a stop. The graph that represents displacement versus time does *not* look like the top one in Figure 12.1 but more like the one at the bottom. The amplitude of the oscillation decreases with time. Friction or air resistance is responsible for this behavior. The larger the friction, the more rapidly the amplitude of oscillation decreases. The energy that was given initially to the oscillator is being slowly changed into some other form of energy. Some of the energy may go into heat. In a musical instrument (piano) some of the energy may be given off as sound energy.

The time taken for the amplitude of an oscillator to decrease to half its original value is called the **damping time** τ (Greek letter tau). Note that the amplitude does not decrease in a straight-line fashion, but rather, the decrease in amplitude becomes smaller as the amplitude itself gets smaller. One way to put it is to say that the loss in amplitude is proportional to the amplitude itself.

Physics in the Arts, Revised Edition

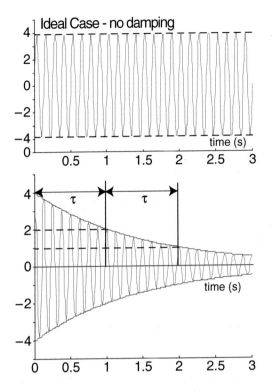

Ideal Case - no damping

FIGURE 12.1 ▼

Two oscillations of the same frequency are shown. If there is no friction or other loss of energy of the oscillator, the amplitude remains constant. But in reality, there is always some loss of energy that makes the amplitude of the oscillations decrease with time, as shown in the lower graph. The "damping time" τ is the length of time it takes the amplitude to decrease to half of the previous value.

As a result, if the amplitude decreases to half in one damping time τ, it will reduce to a quarter in 2τ, as shown in Figure 12.1. This is called an exponential decrease in amplitude. If you inherit a small fortune and every year spend half of what you still own, your capital would decrease in the same way, so you never run out of money! It is the same with a tuning fork: Theoretically it will never stop, but after a few seconds the sound is so soft that nobody can hear it (or, in the case of your inheritance, you are too poor to buy yourself a hot dog!).

A long damping time τ indicates that there is little friction to dissipate the energy of the oscillator, we say that there is little "damping." A short damping time τ indicates that the oscillator is losing energy rapidly. The oscillator is "heavily damped," as in the last graph of Figure 12.2. The vibrations of a drumhead are heavily damped compared to the oscillations of a plucked guitar string, which lingers for several seconds. A 500 Hz guitar string with a damping time of 2 s makes $2 \times 500 = 1000$ oscillations before it loses half the amplitude.

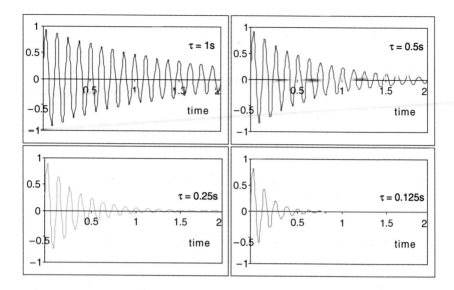

FIGURE 12.2 ▶   Each of the oscillations shown here has a frequency of 8 Hz. The damping time varies from 1 s for the first graph to 0.125 s for the last graph.

12.2   Resonance

How are oscillations in musical instruments set in motion? In many instruments the oscillations are built up by adding many little pushes, timed to correspond to the period of the oscillator, much like a gentle person can set a swing in motion by many properly timed small pushes. As an example, look at a trumpet. The musician, by pressing his lips against the mouthpiece, makes rapid periodic puffs of air (try to "buzz" your lips!) that set into motion the air column in the trumpet. For a trumpet, the buildup of the oscillations is quite rapid, but in a large organ pipe, it may take half a second or more to build up the oscillations.

Another example where an oscillating system is set in motion by repeated pushes from the outside is the soundboard of a piano, which is put in motion by the vibrating string.

Now we will examine the fundamental properties of such "driven oscillators" by first studying simpler systems, such as a pendulum (see Figure 12.3) or a cart between springs, riding on a (nearly frictionless) track (see Figure 12.4).

We will assume that the external force (the **driving force**) is a simple **periodic sinusoidal pull** with a constant frequency. Let's consider the example

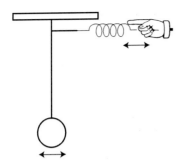

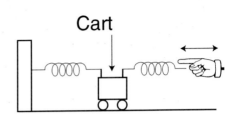

Cart

FIGURE 12.3 ▶ You apply a gentle, periodic force near the top of a pendulum.

FIGURE 12.4 ▶ You apply a gentle, periodic force to a glider on wheels.

of the cart on wheels in Figure 12.4. There are four things to be taken into consideration:

1. The natural frequency f_N (the subscript N is attached to f to indicate that it is the natural frequency). The natural frequency is the frequency of oscillation in the absence of an external driving force.

   In Chapter 11 we learned that f_N depends on the physical construction of the oscillator, such as the mass of the cart and the stiffness of the springs.

   As an example, let us assume that the oscillator, if it is set in motion and then left alone, has a frequency $f_N = 10\,\text{Hz}$.

2. The driving frequency f_D is the frequency with which you pull on the spring. The driving frequency can be whatever one chooses; you are not obligated to push the oscillator at its natural frequency.

   For instance you might pull 5 times per second ($f_D = 5\,\text{Hz}$) or any other frequency.

3. The amplitude of the driving force specifies how hard you are pulling on the spring, such as how large a force you use. You can choose to pull very softly or very hard; a stronger force will result in larger amplitude of the motion.

4. The amplitude of the oscillator measures the response of the oscillator to the driving force.

We want to see how the motion of the glider depends on the driving frequency f_D. We try various frequencies using the *same force* in all cases, and we *observe the amplitude* of the oscillator. Suppose that the oscillator has a natural frequency of 10 Hz, but we start out by pulling on it with a driving frequency $f_D = 4\,\text{Hz}$. Note that f_D is very different from f_N. The glider starts to move a little bit. It wants to move at 10 Hz, but the pushes are at 2 Hz, so the pushes are not at all in step with the natural frequency. In the beginning the motion of the glider is irregular—neither 10 Hz nor 4 Hz—but it is soon forced into vibrating at the frequency of the driving force f_D. However, the amplitude A of the oscillation will be small, as indicated in Figure 12.5.

Now we change f_D to 8 Hz, which is closer to the 10 Hz natural frequency. The first oscillation of the driving push imparts a small amount of motion to the glider. The second back-and-forth push in this case comes 0.125 s after the first push, but this is very close to the 0.1 s that would be needed to be in step with the natural glider motion. This means that for $f_D = 8$ Hz, the glider moves with a larger amplitude than for $f_D = 4$ Hz, where the timing of the pushes was completely off.

To build up the biggest amplitude, the driving frequency f_D should be equal to f_N because then the subsequent pushes keep adding to the motion imparted by the previous pushes. So at $f_D = 10$ Hz, the cart's amplitude will be the largest. When you push with too high a frequency, the amplitude is less again because the pushes are out of step with the natural frequency.

A plot of the cart's amplitude versus driving frequency is shown in Figure 12.5.

The response of an oscillating system to pushes from the outside shown in Figure 12.5 is called "resonance." The oscillating cart resonates at 10 Hz, i.e. it selectively responds to the pushes if they are at or near the resonance frequency.

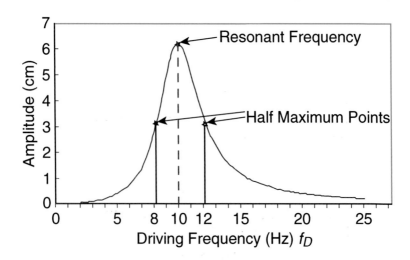

FIGURE 12.5 ▶ Resonance curve for an oscillator with a natural frequency of 10 Hz. In this example, if the driving frequency differs from the natural frequency by 2 Hz either way, the amplitude of the oscillator is still half of the maximum amplitude when you are exactly on resonance. The width of this resonance curve is 2×2 Hz $= 4$ Hz.

For simplicity, we used a pushed cart as an example. What other examples of resonance can you think of? How about singing in the shower? If you have a deep, loud voice and change the frequency from high to low, you may find a big echo at a certain frequency—namely at the natural frequency of the air oscillations in the shower stall volume. If you have access to a piano, one demonstration of resonance you can try is to depress the loud pedal (which takes the dampers off the strings) and sing a loud tone close to the piano. Some string may resonate—namely the strings whose natural frequency is close to the driving frequency of your voice. It may help to hit a key first and then sing the same tone to get your voice's f_D to match the f_N of the string.

How close to the natural frequency (f_N) must the driving frequency f_D be to give reasonably large amplitudes? One way of looking at this is to find at what frequencies the amplitude is exactly half of that obtained when $f_D = f_N$. In Figure 12.5, the maximum amplitude (for $f_D = 10\,\text{Hz}$) is 6 cm. At $f_D = 8\,\text{Hz}$ the amplitude is half the maximum value. The same is true at 12 Hz. The difference between these two frequencies is called the width (Δf) of the resonance curve (also called the "full width at half maximum" FWHM). In this case $\Delta f = 12 - 8 = 4\,\text{Hz}$.

Another resonance curve for a different driven oscillator is shown in Figure 12.6. In this case we would say the resonance frequency is $f_N = 60\,\text{Hz}$ and the width $\Delta f = 20\,\text{Hz}$.

A resonance width of 20 Hz means that if the driving frequency f_D is off by half that amount (10 Hz) on either side of f_N, the response of the oscillator is half the maximum response.

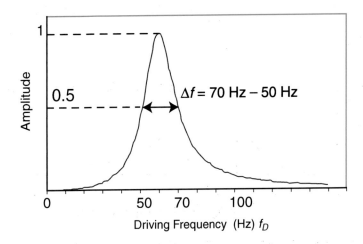

FIGURE 12.6 ▶ Resonance curve of a large Tabla (Indian drum) when set in motion by a periodic force. The resonance frequency f_N is 60 Hz, and the width of the resonance is 20 Hz.

The width of the resonance curve depends on the amount of damping.
One finds that some resonance curves are very sharply peaked, while others
are quite broad. The important difference between sharp or broad resonances
is the amount of damping (amount of friction) of the oscillating system that is
being pushed. If the oscillator has not much damping, one can build up a huge
amplitude, provided the driving frequency is close to the natural frequency.
Why? Because for not much friction (long damping times) the oscillator gets
a little more energy with each push—that is, it still remembers all the previous
pushes. But in this case, even if the driving frequency is off only a little, the
later pushes will eventually have fallen out of step and start to reduce rather
than add to the oscillations.

On the other hand, when the oscillator is damped a lot the energy it
acquired from the pushes is soon dissipated, so the oscillator does not
remember the previous pushes. This has two effects, which are illustrated in
Figure 12.7: (a) for more friction, the peak of the resonance curve is not as
high, but (b) if you are off in frequency it matters relatively less so that the
resonance curve is wider. In summary, there is an inverse relationship between
Δf and τ: *Long damping time (small amount of friction) corresponds to a
small width of the resonance curve, and vice versa.*

Until now, we always assumed that the pushes of the driving force continued for a long time—that the oscillator came to a "steady state." The

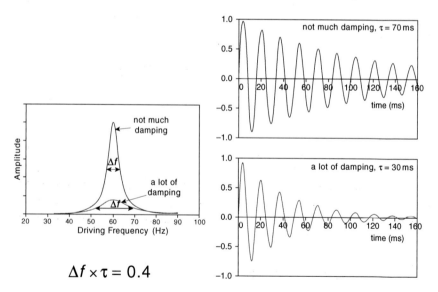

FIGURE 12.7 ▶   The graph on the left shows two resonance curves for different amounts
of damping (friction) but the same resonance frequency. The graphs at right show the
damped oscillations for the same two cases.

resonance curve shows the steady state response to the driving force. The time needed for the oscillations to steady down depends on the damping time: short steadying time for short damping time, long for long damping time.

The inverse relationship between Δf and τ can be put into quantitative terms: A calculation shows that the product of Δf and τ is always the same (at least if the damping is not too large, in which case the product is $\ln 2 \sqrt{3}/\pi$):

$$\boxed{\Delta f \cdot \tau = \text{constant} = 0.4}$$

If you make Δf large, τ becomes small, or if you make τ large, Δf becomes small (long decay and sharp resonance). This relationship has important applications and is extensively used in acoustics.

Exercise 12.1 The drumhead of a kettledrum has a frequency of 50 Hz. The drumhead makes about 10 oscillations before the amplitude has decreased to half the initial value. Find (a) the damping time and (b) the width of the resonance curve.

Answer:

(a) The period is $(1/50)$ s, so 10 oscillations take $(10/50) = 1/5 = 0.2$ s Damping time $\tau = 0.2$ s.

(b) Solving the equation for Δf gives: $\Delta f = 0.4/\tau = (0.4/0.2 \text{ s}) = 2.0$ Hz. Note that the units come out correctly: τ is in seconds, so $1/\tau$ has units 1/s, which is the correct unit for frequency (Hz).

12.3   Build-up and Decay of Musical Tones

When you listen to an instrument (e.g., a clarinet or a violin) you get the impression that the sound of a given note played starts and stops instantly. But if you look at the tone on an oscilloscope, you find that the tone builds up and decays gradually. Buildup and decay times are closely related: If it takes the tone 0.1 s to decay to half, it also takes 0.1 s to build up to half the amplitude. It turns out that the gradual buildup and decay is an important attribute of the tone quality of an instrument. If one records a tone from a musical instrument, and during replay with a fast switch cuts off the build-up and decay, it is much harder to tell what instrument is playing!

12.4   Applications in Music

12.4.1   Resonators in Musical Instruments

Sympathetic strings: Some instruments, including the Indian sitar or the 18-th century viola d'amore, as well as some guitars, use so-called "sympathetic"

FIGURE 12.8 ▶   A modern marimba. The actual resonators are behind the ornamental vertical bars in front.

strings to enhance the sound. These strings are not bowed or plucked themselves but are set in oscillation by a nearby string.

Instruments with resonators. The marimba and the xylophone have sets of wooden bars of different lengths, producing a sequence of tones similar to the piano keyboard. The sound of the bars is reinforced by hollow tubes (resonators) whose dimensions are matched to the pitch of each bar to produce a resonant sound. The origin of the marimba is not known, but in the mythology of the Zulus of South Africa, there is a tale about a goddess called Marimba who made an instrument by hanging gourds below wooden bars struck by a mallet. A traditional marimba from the highlands of Guatemala is shown on the opening page of this chapter. Another instrument, the xylophone, is similar to the marimba, but has a different musical range. A modern marimba is shown in Figure 12.8. The frequency of the struck bars increases from one end of the instrument to the other and the resonators correspondingly should decrease in length across the instrument. It thus seems puzzling that the bars in front of the instrument shown in the figure are symmetric, short in the center and long at both ends. The answer is that the bars in front of many commercial instruments are ornamental only; the real resonators are hidden behind the ornamental ones, and, of course, they follow the laws of physics.

The *vibraphone* is a modern instrument (often used in jazz) that we mention here because it provides a vivid demonstration of resonance. The instrument has a set of metal bars and metal tubes of different lengths below each bar. Above each resonator is a metal disc, slightly smaller in diameter. These discs are attached to a motor-driven horizontal bar, such that the resonators are periodically covered and uncovered, causing the resonators to be rapidly turned on and off. Correspondingly, the sound intensity is modulated rapidly from loud to soft ("wow-wow-wow"). The musician can change the speed of the motor to change the rate of modulation.

Soundboards of instruments: The sound volume produced by a piano string by itself is very weak because the string itself is small and does not push on much air. The beautiful, loud sound of a good piano (or any other string instrument) is achieved by running the taught string over a bridge that transmits the vibrations to a large *sound board*. Some people speak of a "resonance board," but this is poor terminology because the instrument builder wants to avoid resonances. Why? Because a resonance selectively responds to a narrow band of frequencies, while the sound board should respond evenly to all frequencies of the instrument.

A related comment can be made about loudspeakers: Most *loudspeakers* consist of a cardboard cone with a corrugated flexible rim that allows the speaker cone to move in and out. The driving force is provided by an electrical coil and a magnet that pushes on the apex of the speaker in step with the electrical signal from the amplifier. So, indeed, this is an oscillating system with a periodic driving force. To achieve a more or less uniform frequency response, we want the resonator to be very heavily damped so that the resonance is flat and spread out. To obtain good music reproduction over all frequencies, one uses not just one speaker but separate speakers for bass (*woofer*), midrange, and high frequencies (*tweeter*). An unavoidable consequence of the need for heavy damping (high friction) is that most of the energy supplied by the amplifier goes into overcoming the friction. A 50 watt music system produces but a small fraction of 50 watts as sound energy; most energy produces heat instead of sound.

While not related to music, *touchtone dialing* (technically called dual-tone, multifrequency dialing) is an interesting everyday application of resonance. When you press one of the 12 keys on a telephone keyboard, you hear two pure tones simultaneously. Seven tones are generated by your telephone and are sent to the telephone switchboard center two at a time. At the switchboard, the signal is sent to seven resonance circuits, each of them responding to one particular frequency. Seven tones are sufficient to represent the 12 keys. For instance, when you push 1, the two frequencies are 697 Hz and 1209 Hz. It is said that some people with good tone perception can tell the number from the tones.

12.5 Questions

1. The motion of a tuning fork is damped.
   (a) If the damping time is 1.5 sec, what fraction of the initial amplitude remains after 3 s? After 6 s?
   (b) Is the frequency of the fork the same after 3 sec as it was at the beginning, or is it lower? Higher? What is the implication of the answer as far as the pitch of the tone is concerned; in other words, how does the pitch of a guitar or a tuning fork change as the tone dies away?

2. *Damping time:* A guitar string has a frequency of 260 Hz (middle C) and a damping time of 2 sec. How many oscillations does the string make before the amplitude has decreased to half of the original value? If the initial amplitude of the guitar string was 2 mm, how large will the amplitude be after 2 sec? After 4 sec? After 8 sec?

3. The following graph shows the resonance curve of an oscillating
system—for instance, a flexible metal blade or the air in a tuning
fork box.

(a) What quantity does the *horizontal* axis of the graph represent?

(b) What quantity does the *vertical* axis of the graph represent?

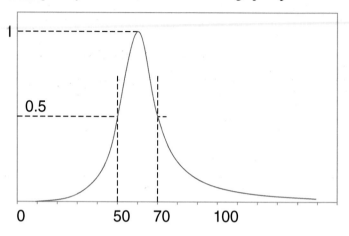

(c) From your understanding of the figure, what is the natural
frequency of the oscillating system? How large is the width of the
resonance curve?

(d) From the preceding answers, deduce the *damping time* of the
oscillating system.

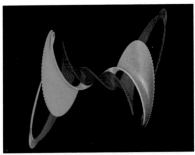

Adding Sound Sources: Beats and Harmony

13.1 Principle of Superposition

Sound consists of rapid pressure variations. If the pressure variations are periodic, we perceive the sound as a tone, not as noise, which consists of disordered (random) pressure variations.

What happens to the pressure variations at your ear (or at a microphone that picks up the sound) when two or more sound sources oscillate at the same time—for instance, an entire brass band all playing together?

The way to look at this problem is to first consider that each instrument, if it played by itself, would produce a certain sound pressure variation with time. Let us call the sound pressure of instrument #1 at a given instant in time $P_1(t)$. Then let us have instrument #2 play and call its sound pressure $P_2(t)$. What is the sound pressure when both instruments play at the same time? The *Principle of Superposition* says that the new pressure is the sum.

Pressure created by many instruments = sum of the pressures created by each instrument:

$$P(t) = P_1(t) + P_2(t) + P_3(t) + \cdots$$

Example: One flute produces the pressure variations shown as the black curve in Figure 13.1, while the second flute produces the pressure variations

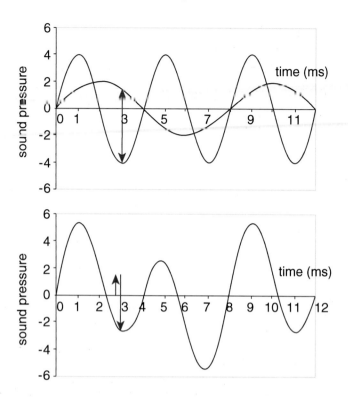

FIGURE 13.1 ▶   Sound pressure variations from two flutes. At time $t = 3$ ms, one flute produces a positive pressure of 1.4 units, the other a negative pressure of −4.0 units. The lower graph shows the sum (or ***superposition***) of the two sounds.

shown as the blue curve. When both flutes play, the pressure variations are the sum (bottom graph in Figure 13.1). Note that in adding the pressures we have to take into account that at a certain instant one instrument may produce a negative pressure (pull on the eardrum), while the other produces a positive pressure (push). At the instant marked in the graph by the red arrows ($t = 3$ ms), $P_1 = +1.4$, $P_2 = -4.0$, so the sum at that moment is $P = -2.6$. At $t = 4$ ms both pressures are zero, so the sum is zero.

We now discuss superposition for *some important special cases*.

13.2   Two Pure Tones of the Same Frequency

"Pure tones" correspond to sinusoidal pressure variations, such as tuning fork oscillations. If one adds two sine curves of the same frequency, the result is again a sine curve with the same frequency, but the amplitude of the sum depends on the relative timing of the two oscillations. To understand this, imagine two swings of the same length side by side: They can be set in motion in synchronism (in step), or they might be swinging out of step with one another. The "out-of-stepness" is measured by a quantity called the phase shift φ (Greek letter phi), where one full oscillation corresponds to 360° phase shift. Thus, for example, a 90° phase shift corresponds to 1/4 oscillation, as shown in Figure 13.2. You will note in the Figure that the sum is again a sinusoidal oscillation, but the amplitude is not simply the sum of the two amplitudes because the two oscillations do not reach the peak at the same time.

If the two oscillations are shifted by 180°, they completely oppose each other. In this case when one oscillation is at its peak ($P_1 = 5$) the other is at its trough ($P_2 = -3$), and the amplitude of the sum is only two pressure units—that is, less than either instrument alone! This is shown in Figure 13.3. Therefore, instead of being louder, the two sounds together are softer than either one alone. Could this really happen? Yes, if you produced the tones electronically and fed them to the same speaker—but not for a musical instrument. Musicians cannot

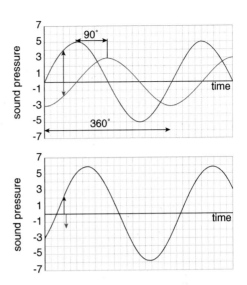

FIGURE 13.2 ► Two sinusoidal oscillations of the same frequency but different amplitudes (5 and 3 units of pressure) are added. The oscillations are not in step but are shifted in phase by 1/4 of a full oscillation. If we consider a full oscillation to represent 360°, the phase shift here is 1/4 of 360° or 90°. The Figure on the bottom shows the sum of the two oscillations.

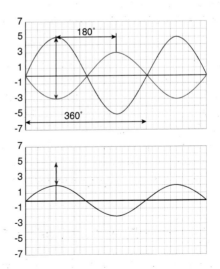

FIGURE 13.3 ► Same oscillations as in Figure 13.2 but with 180° phase shift.

time their instruments so accurately as to remain exactly out of phase for any length of time.

How does one measure the phase shift? A simple way is to count how many divisions on the graph you would have to shift one of the curves to get the two curves in phase. Next, you count how many divisions one period corresponds to and call this 360° (as in Figures 13.2 and 13.3). Now you use simple proportions to find the phase shift in degrees.

Example: In Figure 13.2 the red oscillation is shifted to the right (delayed in time) by 5 divisions with respect to the blue oscillation. One full oscillation is 20 divisions $= 360°$. Thus, 1 division $= 360°/20 = 18°$, and 5 divisions $= 5 \times 18° = 90°$.

13.3   Beats

Consider two tones of nearly the same frequency—for instance $f_1 = 100$ Hz and $f_2 = 98$ Hz. For simplicity we assume that both tones have the same amplitude. In musical terms, the two musicians play the same tone but are slightly out of tune, as they might be before the performance when their instruments are not well tuned.

In this case, if the two oscillations are initially in phase (in step), they will slowly go out of phase because one frequency is a little higher than the other, which means that one period is a little longer than the other. In this example, after 1/4 second, the first instrument (100 Hz) has made 25 oscillations, while the second instrument (98 Hz) has $98/4 = 24.5$ oscillations. This means the corresponding oscillations are shifted with respect to each other by 0.5 oscillations, which corresponds to a phase shift of 180°. They now are opposing each other, and now the two sounds cancel.

Figure 13.4 should help to visualize the situation. At the time 0 there is a maximum for both oscillations #1 and #2. As time goes by, the maxima of oscillation #1 occur before those of #2. Eventually, the maxima of oscillation #2 arrive so late that they coincide with the minima of oscillation #1 (position A on the time line of Figure 13.4). They are now 180° out of phase. As we carry on the drawing, the maxima of oscillation #2 lag further and further behind until one full cycle is missed, and a maximum in #2 happens at the same time as the "next" maximum of #1 (position B). The two oscillations are 360° out of phase, but that is the same as being in phase again!

The term *beats* refers to the acoustic effect of alternating loud and soft sound coming from the alternating in-step and out-of-step oscillations. The *beat frequency* is the repetition rate of loud and soft alternations—that is, how many times a second the two sounds cancel.

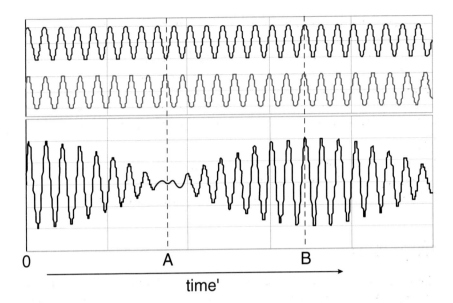

FIGURE 13.4 ▶ Oscillations #1 and #2 with slightly different frequencies. During the time the first oscillation (blue) makes 17 full cycles, and the second one (magenta) makes 16. The sum of the two oscillations (bottom) shows "beats."

In general, if one adds two oscillations with frequencies f_1 and f_2 and corresponding amplitudes A_1 and A_2, these *add* most strongly $(f_1 - f_2)$ times per second. At these times the resulting amplitude will be $A_1 + A_2$. Also they subtract from each other $(f_1 - f_2)$ times per second, and at these times the resultant amplitude will be $A_1 - A_2$ (or $A_2 - A_1$, whichever is positive).

The beat frequency is equal to the difference in the frequencies of the two sources:

Beat Frequency:

$$f_B = (f_1 - f_2)$$

Beats between piano strings: If you open up a piano you will notice that the same hammer strikes two (or on a grand piano even three) strings. When a piano is badly out of tune, there will be rapid beats between the two strings producing the rough, unpleasant sound normally associated with a honky-tonk barroom piano. On the other hand, in tuning a concert grand piano, the tuner may well choose to allow a very small mistuning so that the sound that lingers has a soft, long modulation in amplitude.

13.4   Harmony

Certain tones played together sound pleasant to us—harmonious—while others sound rough—dissonant. This is not to say that dissonant sounds are not used in music. It is often the contrast between harmony and dissonance that makes a composition interesting.

If you hit two keys on the piano at the same time, certain combinations sound "in tune" or harmonious. These combinations are given special names, depending on how many keys apart they are, as we will see later when we talk about musical scales. For now we just want to mention a few of them and explain what is special about harmonious intervals.

The frequency relationship between two tones is called a *musical interval*. An interval is characterized by the *frequency ratio* of the two tones. Some harmonious intervals ("consonant" intervals as opposed to "dissonant" intervals) are listed here:

Name of Interval	Frequency Ratio	Example
unison	1:1	300 Hz + 300 Hz
octave	1:2	300 Hz + 600 Hz
fifth	2:3	200 Hz + 300 Hz
fourth	3:4	150 Hz + 200 Hz

You note that the *frequency ratios* in all these cases are what you might call *simple-number ratios*, like 2:3 rather than for instance 23:31. The superposition of two tones of ratios like those shown in the preceding chart has a special property: The sum is periodic and repeats itself after a short time. You can look at it this way: For a frequency ratio—say, 2:3—in the same time interval in which the 200 Hz tone does two full oscillations, the 300 Hz tone does three full oscillations, as illustrated in Figure 13.5. Notice that the superposition of the two tones is nonsinusoidal but periodic with a period of 10 ms. Why 10 ms? The periods of the two tones (200 Hz and 300 Hz) are 5 ms and 3.33 ms, and we must wait long enough for *both* to have completed an integer number of oscillations. After 10 ms, the two oscillations have completed two and three full oscillations, respectively. The repetition period is thus 10 ms $= 0.01$ s, and the repetition frequency is $f = 1/T = 100$ Hz.

In general, the frequency with which the superposition tone repeats itself is the *largest common divisor* of the two individual frequencies. The largest common divisor of 200 and 300 is 100. (A number like 50 would be a common divisor also, but 100 is the *largest* one.)

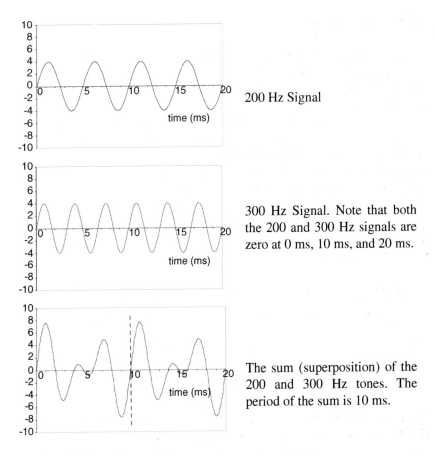

200 Hz Signal

300 Hz Signal. Note that both the 200 and 300 Hz signals are zero at 0 ms, 10 ms, and 20 ms.

The sum (superposition) of the 200 and 300 Hz tones. The period of the sum is 10 ms.

FIGURE 13.5 ▶ Superposition of tones with frequency ratio 2:3.

13.5 For the Fun of It: Lissajous Figures

Jules-Antoine Lissajous had the idea to graph two oscillations in a different way: show the displacement that corresponds to one oscillation along the x-axis, while the other displacement is shown along y. You can see a simple demonstration of this in the Exploratorium in San Francisco, where a pendulum like that in Figure 13.6 hangs from the ceiling. The pendulum has a different period for x and y motion. The curve traced out by the paint bucket (or by a felt pen) is a Lissajous figure.

What do we need Lissajous figures for? We don't really need them, but they are fun to physicists and perhaps also to some graphic artists (see Figure next to chapter title). Furthermore, as we shall see shortly, they can be used in the lab to tune two tones to be exactly in a harmonious ratio. In this case we

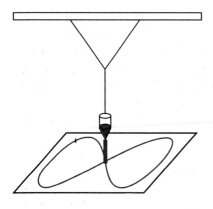

FIGURE 13.6 ▼

A Lissajous pendulum. The pendulum swings left to right faster than back and forth because it is shorter in one direction than the other. If you make it oscillate left to right und back and forth at the same time, the blue paint dripping down from the paint bucket traces out a curve called a Lissajous figure.

would use an oscilloscope and connect one signal to the x input, the other to the y input so that the dot on the face of the scope gets deflected accordingly. Let us look at some examples:

Same frequency: This would correspond to the oscillations of a normal pendulum (without the V in the string of Figure 13.6), which oscillates along x or along y or any other direction with the same frequency.

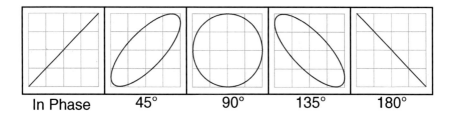

In Phase   45°   90°   135°   180°

FIGURE 13.7 ▶   Lissajous figures for the frequency ratio 1:1.

The graph you get depends on the phase shift between the x and y oscillations. Think about it: If the x and y oscillations are in step, the pendulum traces a straight line. If both amplitudes are the same, the straight line is at 45°, as shown in Figure 13.7. Now if you imagine the two oscillations being out of phase by 90° (a quarter of an oscillation) as in Figure 13.2, you see immediately that when one displacement is at the maximum the other is zero: The pendulum moves in a circle if x- and y-amplitudes are the same. For phase shifts intermediate between 0° and 90° the path of the pendulum is an ellipse. For 180° phase shift it is again a straight line. Note that all of these are curves that come back upon themselves after one oscillation, because the period in x and y displacements is the same.

Frequency ratio 2:1: If one oscillation is twice as fast as the other (octave interval in music), the pictures again depend on the relative timing of the two oscillations—that is, if they cross zero at the *same* time as was shown in Figure 13.1 or are shifted with respect to one another. In contrast to Figure 13.1, here we assume again that the amplitudes in x and y are the same so that the Figure fits into a square. Try to figure out which of the graphs in Figure 13.8 corresponds to the relative timing of the oscillations in Figure 13.1.

Whether you think this is fun is a matter of taste. But it is good to understand that *Lissajous figures are very useful for exact tuning to a simple frequency ratio*. If the frequency ratio differs ever so slightly from the exact ratio, the figure does not repeat itself again and again but rather changes slowly from one of the patterns to the next.

What is the frequency ratio? Try to figure out on your own what the frequency ratio is for the set of pictures shown in Figure 13.10. You can get a hint if you ask yourself why the curve for the 2:1 ratio touches the top and bottom of the graph twice while it touches the left and right side only once.

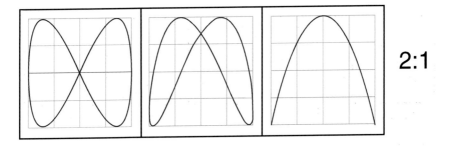

2:1

FIGURE 13.8 ► Lissajous figures for 2:1 frequency ratio.

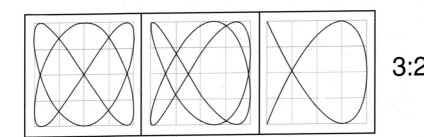

3:2

FIGURE 13.9 ► Lissajous figures for 3:2 frequency ratio.

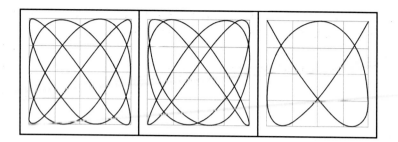

FIGURE 13.10 ▶ What are the frequency ratios for each of the three Lissajous figures shown here?

Nonsimple ratios: For a frequency ratio like 1:1.571 almost the entire square will be filled with paint before the line closes. You may remember from math that some numbers cannot be expressed as a ratio of two whole numbers. For such "irrational" relationships between the two frequencies, the curve never closes; the entire page will eventually get filled with paint.

13.6   Questions

1. Add the two oscillations below between 0 and 1.0 ms (one point at every 0.2 ms mark and one point in between will be sufficient)
   What is the frequency of oscillation A?
   What is the frequency of oscillation B?
   What is the frequency of the sum of A and B?

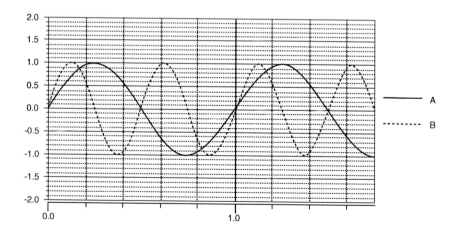

2. What is the physical basis of harmony—in other words, why do only certain frequency combinations sound harmonious?

3. *Beats:* A tuning fork has a frequency of 440 Hz. What are the possible frequencies of a second tuning fork if we hear 4 beats every second when both forks are sounded together?

4. Find the phase shift between the two oscillations shown in the following graph.

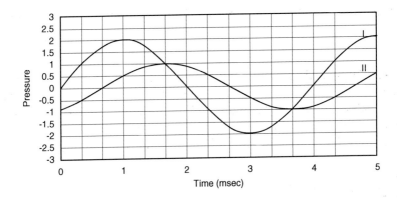

▶ Chapter 14

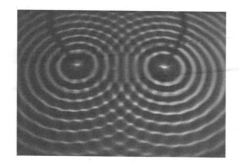

Sound Waves

14.1 Propagation of a Pulse

If you drop a stone in a still pond or an undisturbed swimming pool, a disturbance is created on the surface of the water. What happens is shown in Figure 14.1. One sees that the disturbance, or bump, on the water surface travels across the pond.

Speed of propagation: One can measure the speed with which the bump travels across the water. To find this speed, divide the distance traveled (D in Figure 14.1) by the time ($t_2 - t_1$) it took to travel that far:

$$v = \frac{D}{(t_2 - t_1)}$$

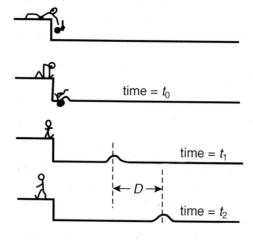

FIGURE 14.1 ▶ A rock is dropped in a pond.

If you put some leaves or paper boats on the water, you see that they bob up and down as the disturbance moves under them. They ***do not*** move along the surface with the bump, but the water moves only up and down. Energy of motion is being transmitted across the pond by the water without any water actually moving from one side of the pond to the

Physics in the Arts, Revised Edition
© 2012 Elsevier Inc. All rights reserved.

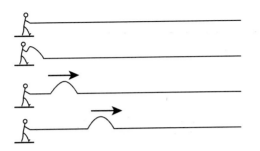

FIGURE 14.2 ▶ Transverse pulse on a string or a rope.

other. The same with sound propagating in air: The air molecules push on one another and transmit the sound from place to place without any motion of air from the instrument to the listener.

The same phenomenon of a traveling disturbance can be seen on a stretched Slinky or on a taught, long rope or string: Shake one end sharply, and you can see a bump traveling along it. Exactly the same happens on the taught string of a piano or a string instrument. To understand vibrations of strings, we need to figure out the speed of propagation (see Figure 14.2).

The height of the bump obviously depends on how strongly one has shaken the end of the rope. The speed of propagation of the bump does *not* depend on *how* the end was shaken but only on the **tension** on the string and on **mass per unit length** of the string

Speed of propagation along a taught string $\quad v = \sqrt{\dfrac{F}{M/\ell}} = \sqrt{\dfrac{F}{\rho}}$

Here F is the tension on the string (in Newtons)—that is, the force applied at the ends of the string to keep it taught—M is the mass of the string (in kg), and ℓ is the length of the string (in m). What matters to the speed of motion is the quantity M/ℓ—that is, the **mass of 1 meter of string**. Whether you cut a long or a short piece of string to figure out M/ℓ does not matter because M and ℓ increase in direct proportion to one another. So sometimes it may be better to call M/ℓ by a new name, like Greek rho: $\rho = M/\ell$ = mass per meter length (also called "linear mass density") of the string. Usually, the symbol T is used for the tension, but to avoid confusion with the period T, we simply call it F.

Does the formula for the speed of propagation make sense? To understand the formula, we look at the propagation of a pulse in detail. For our string we will use a large number of balls strung like beads onto a strong thread. Then we stretch the thread with a tension, F, hit one ball with a hammer, and watch the pulse propagate as in Figure 14.3.

Intuition will tell you that when the first ball is hit downward by the hammer, the second ball also moves downward but a little *later* than the first ball. Why? Because after the first ball has moved a small distance, the taught string starts pulling on the second, but it takes time for the second ball to pick up speed.

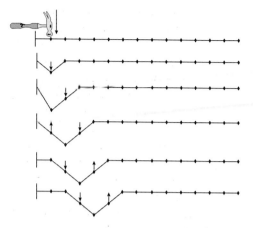

FIGURE 14.3 ► The dots represent balls on a taught, stretched string. The small arrows indicate the direction of motion of the balls.

For the same reason, the third ball will move only after the second ball moves and with the same delay. The accumulation of delays, combined with the separation of balls, yields a certain speed for the pulse propagation. This speed is higher if the *delay* is smaller and vice versa. Since heavy balls move "sluggishly" for a given force, the delay is longer for heavy balls than it is for light balls. Similarly, if the balls are closer to each other, there is more mass to be set in motion and the propagation is slower. So the **greater the mass per meter length,** M/ℓ, **the slower the pulse** travels. It thus makes sense that M/ℓ in the formula is in the denominator. If we increase the tension, F, the force on the second ball due to the first is larger, it will pick up speed (accelerate) faster and the delay is shorter. So **greater tension F means faster pulse speed v.** Again the formula works out. The square root sign is harder to understand intuitively, but without the square root, the units do not work out—just as we saw in Chapter 11 for simple harmonic motion.

To summarize:

- *The greater the tension, the higher the speed of propagation.*

- *The heavier the rope or the string, the slower the speed of propagation.*

A word of caution: The speed of propagation formula is true only if the string is very limber—that is, if it has no stiffness. Any stiffness of the string adds an additional force (besides the tension) that is not taken into account in the formula. It should be emphasized that the speed of sound propagation depends on the medium. The formula given here applies only to limber, taught strings, and not for instance to sound waves in air.

14.2 Longitudinal and Transverse Waves

If one end of the rope (or of a Slinky) is moved smoothly up and down at a constant rate, a *wave* travels along the string. You can do this by stretching a long Slinky on the floor and moving one end smoothly back and forth again

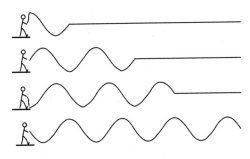

FIGURE 14.4 ▶ Transverse sinusoidal wave on a long string.

and again. Figure 14.4 shows the shape of the Slinky at different instances in time. A wave is different from a simple oscillation: A wave extends in space with many points along the wave oscillating at the same rate. The waves we discuss here are mechanical waves, quite different from light waves that, as we learned, are electromagnetic waves.

Note that in the Figure 14.4, the wave travels from left to right, but each point of the string moves up and down; that is, the motion of the string is at 90° from the direction of propagation. This type of wave is called a transverse wave.

Now suppose you move the end of a Slinky back and forth *along* the Slinky axis rather than transverse to it—in other words, you push and pull on the Slinky again and again to compress and to stretch it periodically. Now the direction of motion of each part of the Slinky is the same as the propagation direction (Figure 14.5): This is called a longitudinal wave.

14.3   Sound Waves in Air Are Longitudinal Waves

Air is squishy. It can be compressed, as you well know if you ever used a simple bicycle pump. Sound waves in air are longitudinal waves. A big loudspeaker membrane that moves rapidly back and forth making alternate compressions and decompressions can create sound waves. When compressions are passed to the neighboring air particles, they are set in motion because the compressed air near the speaker pushes on the nearby air particles, much like we visualized for longitudinal waves on a Slinky.

In Figure 14.6, the dots symbolize the air molecules. At a wave crest the air is denser; at a wave trough the air is more dilute. You should imagine the crests and troughs to be moving to the right, assuming a loudspeaker continues to produce compressions at the left. The picture should not be taken too literally. Obviously, there are many more molecules than we can show, but also the change in air density from one place to the next is tremendously exaggerated. Even for the very loudest sounds you can barely tolerate, the change in air density between the wave crests and troughs is only 1/10,000 of the normal air pressure! So if normal air pressure (no sound) had 10,000 little dots on one area of Figure 14.6, at the peaks there would be 10,001 dots and at the troughs

PULSE

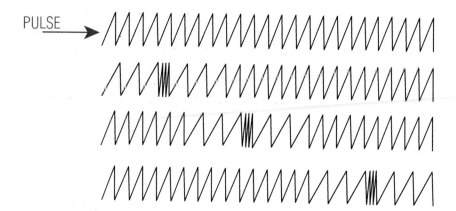

FIGURE 14.5 ▶ Longitudinal pulse on a Slinky.

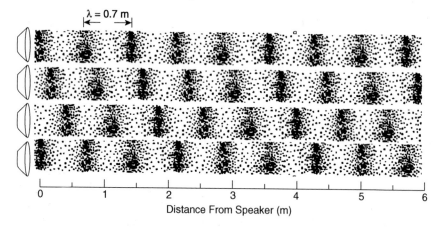

FIGURE 14.6 ▶ Longitudinal sound wave from a loudspeaker. The wave moves left to right and is shown at different instances in time. Greater density of dots indicates greater air pressure. The distance from one wave crest (maximum density) to the next is called the wavelength λ The bottom drawing is lagging one period T behind the top drawing.

9,999 dots. You certainly would not be able to notice it. Your eardrum is just terribly sensitive: An ant walking on your eardrum would make a deafening sound!

The distance from one wave crest to the next is called the wavelength λ (Greek letter lambda). The wavelength is marked in Figure 14.6. We encountered the wavelength concept already in Chapter 1, but for audible sound the wavelengths turn out to have everyday dimensions from inches to yards, while for visible light the wavelengths are less than a nm.

Making a drawing like Figure 14.6 full of little dots has two shortcomings. It is tedious to draw, and it is not *quantitative* (i.e., it is difficult to read or measure how compressed or rarefied the air is at a particular point in space and time).

A more convenient way of depicting the sound wave is to draw a series of graphs showing the pressure at various points for different instants in time (Figure 14.7). One sees that the pressure variations are sinusoidal and move to the right. The drawing in Figure 14.7 corresponds to the propagation of a pure tone of about 500 Hz frequency in air.

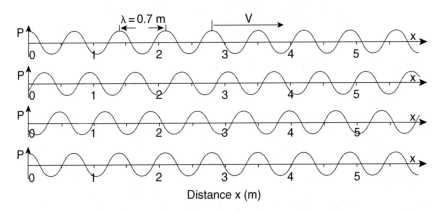

Distance x (m)

FIGURE 14.7 ▶ Graphs corresponding to the pressure distribution of the drawing in Figure 14.6. With time the wave crests move more and more to the right.

Why can't you produce transverse waves in air? If you were to hold a stiff piece of cardboard in front of you and move it left and right, it would simply slide through the air without making a wave because the air particles are not attached to one another as they are in a Slinky. But if you were to move the same cardboard fast enough toward and away from you, the air would be compressed and rarefied, and a longitudinal wave would propagate.

14.4 Speed of Sound in Air

To understand wind instruments of all kinds, we need to know the speed with which the pressure waves propagate in the instruments. The speed of propagation depends on the properties of the gas. If the molecules of the gas are relatively heavy, they move more slowly than for lighter molecules. Consequently, it takes longer for the compression to reach the neighboring molecules, and thus sound propagates slowly. For example, in helium, whose atoms are much lighter than air (which is made up of mostly nitrogen and oxygen molecules), the speed of sound is almost three times greater than in air.

With increasing temperature the gas molecules move more rapidly, and the speed of sound increases accordingly. The kinetic energy of the gas molecules, $(1/2) mv^2$, is proportional to the absolute temperature. Absolute zero temperature is at $-273.2°C$ ($-460°F$), and at this temperature all molecular motion stops. To figure out the change in speed of sound for instance between 0°C (freezing point of water) and 20°C (room temperature, 68°F), you would add 273.2°K to get the absolute temperatures: 273.2°K and 293.2°K. Now the ratio between these two numbers is 0.932. This is the ratio of molecular kinetic energy—that is, the ratio of the square of velocity at the two temperatures. To find the ratio of velocities, you need to take the square root of 0.932, which is 0.965. So if the speed of sound at 20°C is 344 m/s, at 0°C it is 0.965 × 344 m/s = 332 m/s.

To save you the trouble of doing this calculation for different temperatures, it is sufficient to know two numbers: the speed of sound in air at room temperature and the change in speed of sound for every degree change in temperature.

> *speed of sound in air at room temperature (20°C) v = 344 m/s*
> *increase in v with temperature: 0.17% per °C = 0.6 m/s per °C.*

In U.S. customary units (speed in feet/s and temperature in °F), the corresponding numbers are:

> *speed of sound in air at room temperature (68°F) v = 1130 ft/s*
> *increase in v with temperature: 1.1 ft/s per °F.*

The change of speed of sound with air temperature is important for brass and woodwind musicians. The pitch (frequency) of the instruments depends on the speed of sound and, therefore, depends on the temperature of the air inside the instrument. As we will see later when we talk about oscillations of wind instruments, the frequency changes in direct proportion to the speed of sound. A wind instrument gets warmer after it is played a few minutes, so when a flutist warms up her instrument—for instance, by 10°C (18°F)—the speed of sound increases by 10 × 0.17% = 1.7%. This raises the pitch very noticeably.

14.5 Wavelength and Frequency

The wavelength is the distance from one wave crest to the next, as shown in Figure 14.7. But you could just as well have used the distance from one wave trough to the next, or the distance from any one point of the wave to the next point, which has the same phase. A wavelength simply includes one complete oscillation.

The rate at which the wave crests arrive at the listener depends on two things: the *speed v* with which the wave travels toward the listener and the *wavelength* λ. You will easily be able to figure out the relationship between frequency and wavelength. If the sound source makes f oscillations in one second, and the wave travels v meters away from the source in one second, there will be f oscillations distributed over v meters, so the wavelength must be

$$\boxed{\textbf{wavelength}: \lambda = v/f}$$

Note that the units in the preceding equation work out: v has units of m/s, and f has units of 1/s, so v/f has units of m. The relationship applies to all kinds of waves: waves on strings, sound waves in air, even light waves, provided one uses the speed v that is appropriate to the particular medium. The equation says that if the oscillation frequency is high, the wavelength is small. Well, yes, *if* you move your big toe in the bathtub back and forth rapidly, the wave crests are close together, because you make more ripples than if you moved the toe slowly. Instead of memorizing this equation, you can always figure it out quickly.

Example: Find the wavelength in air of a very low-pitched sound (100 Hz) and a high-pitched sound (10,000 Hz = 10 kHz).

Answer:

For 100 Hz: $\lambda = \dfrac{v}{f} = \dfrac{344\,\text{m/s}}{100/\text{s}} = 3.44\,\text{m}$

For 10,000 Hz: $\lambda = \dfrac{v}{f} = \dfrac{344\,\text{m/s}}{10,000/\text{s}} = 0.034\,\text{m}$

In U.S. customary units, for a low-pitched tone, the wavelength is about 10 feet, while for a high-pitched tone the wavelength is just over an inch.

14.5.1 Relevance to Size of Instruments and Loudspeakers

The preceding example shows that bass tones have a long wavelength. To produce a strong bass tone, the instrument or loudspeaker must be large. A vibrating violin string by itself will hardly make any sound because as it vibrates back and forth when air is pushed away as the strings move one way, the air simply goes around to fill the void on the back of the string. You can best see this if you hold a pencil vertically in the water and move it back and forth. For a slow motion the water has time to flow around the pencil

and makes hardly any waves; only for a very rapid motion, when the waves
have a very short wavelength, does a pencil make noticeable waves. To make
waves with a long wavelength, you need a large object, like a canoe paddle!
The conclusion: The size of the sound source (the "radiator") must not be
much smaller than the wavelength of sound that is to be radiated. There is an
English saying: "You can't fan a fire with a knitting needle." Thus, a bass fiddle
or a bass speaker needs to be fairly large. A violin radiates higher frequencies and thus can be smaller. Bass speakers increase the efficiency of sound
radiation at low frequency by suitable enclosures that prevent air that is pushed
away at the front of the speaker from going around to the back of the speaker
and filling the void in the back rather than projecting the sound out of the
front.

14.6   Sound Propagation

1. **Reflections**: Sound waves disperse through rooms and to adjacent rooms
   by multiple reflections. Even if there are obstructions between you and
   your radio speaker, you hear the sound reflected from the walls or the
   ceiling (Figure 14.8).

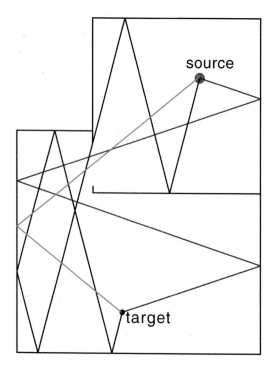

source

target

FIGURE 14.8 ▼
Three possible paths from a
source of sound to a target (a
person) are shown. The three
paths shown are not all possible paths. Shown are paths
with 1, 3, and 6 reflections,
respectively.

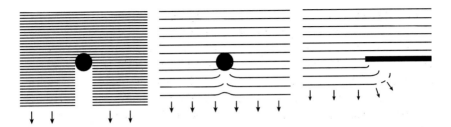

FIGURE 14.9 ▶ Wave propagation near an obstacle. If the wavelength is much shorter than the size of the obstacle, a shadow is created (left figure). The larger the wavelength compared to the size of the obstacle, the more the waves bend around the obstruction (diffraction of waves, pictures on the right).

2. *Diffraction and shadowing*: No doubt you have noticed that another person standing in front of a table lamp makes a shadow on the book you are reading but does not obstruct the sound waves from your table radio. Why are light waves more easily obstructed than sound waves?

The answer is that light waves have very short wavelengths, while sound has wavelengths ranging from a few inches to a few feet.

Shadowing of sound or light waves depends on the size of the object compared to the wavelength. If the wavelength is short compared to the obstacle, there is a shadow behind the obstruction (Figure 14.9). This is the usual case for light waves because the wavelength of light is so short that normal objects are always much larger than the wavelength of light.

But sound waves of lower frequencies have rather large wavelengths, which bend around the obstacle (Figure 14.9).

This would suggest that if you sit behind a big stone pillar at a concert in a Gothic church, the high frequencies would be completely shadowed. If the pillar has a diameter of 1 m, wavelengths much shorter than 1 m would be shadowed. But a 1 m wavelength corresponds to 344 Hz—that is, much above 344 Hz would not be heard. So almost half of the tones on a piano keyboard would not be heard. In reality, even if you sat behind the pillar, where you get very cheap tickets, you still hear the higher frequencies as reflections from the many other surfaces where sound is reflected (Figure 14.8).

14.7 Interference of Sound Waves

The term *interference* of sound waves is commonly used to describe the superposition of sound waves of the same frequency from two sources at different distances from the listener. The term is misleading because the oscillations caused by different sound sources do not really hinder one another, as we see with water waves in the figure on the chapter opening page. The two waves simply add as we have learned earlier when we discussed superposition.

As an interesting illustration we can imagine stereo loudspeakers that are placed some distance apart, as shown in Figure 14.10. Suppose that your amplifier is switched to mono, which means that both speakers are connected to exactly the same signal and oscillate in phase (in step). What will be the result of the superposition of the two sound sources? At first you might say that they will simply add to one another, since they are in step. But this is not the case; we must take into account that we may be closer to one speaker than the other so that the wave from one speaker arrives a little later than the wave from the other one. You might say that sound travels pretty fast, so the delay would not amount to much, but on the other hand, sound oscillations are pretty fast, so the delay should always be taken into account.

What will happen? To think of a specific example, assume that both speakers emit the same pure tone of 1100 Hz, for which the wavelength is about 0.3 m. If you are the same distance from both speakers, the waves arrive in step (in phase) because the delay from the speaker to your ear is the same for both speakers. But if the distance differs, then one wave will be delayed with respect to other. If the delay is just half an oscillation, the two waves tend to cancel; they make "destructive interference," as shown in the bottom picture

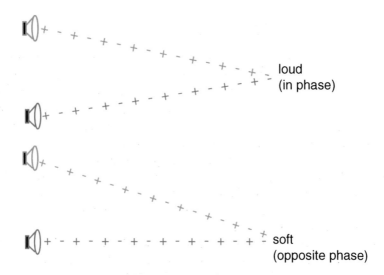

FIGURE 14.10 ▶ Interference from two loudspeakers. The + and − signs indicate wave crests and wave troughs. When + meets + and − meets −, the sounds reinforce (as shown in Figure 1.2). In the figure at the bottom, the listener moved to the side. Now the distance to the two speakers is no longer the same. If the difference in distance equals 1/2 or 3/2 or 5/2 wavelengths, + meets −, and the sounds tend to cancel one another.

of Figure 14.10. This is the case when the difference in distances from the two sources is exactly one-half wavelength—in our example 0.15 m. But come to think of it, what if you move your ear more to the side so that the distance from the two speakers is bigger? Well, when the delay is one whole wavelength the two waves will reinforce again just as in the middle (we have constructive interference). Thus, you can well imagine that as you move off to the side, there will be alternate places of loud and soft sound—loud every time the difference in distance to the speakers is 1 or 2 or 3 whole wavelengths, and soft whenever the difference is 1/2, 3/2, 5/2, and so on, wavelengths.

In a room the effects are not so pronounced because there is also sound coming from wall reflections. But still there is a practical application to test whether your stereo speakers are correctly connected. For mono the speakers are supposed to oscillate together in phase, but if you have one pair of wires reversed, they are 180° out of phase. To do a test, put one of your ears near the speakers the same distance from both speakers, and cover your other ear to reduce hearing room reflection. If you have a friend reverse the pair of wires on one speaker while you listen, the sound will be loud for the correct connection and soft for the incorrect one, since for the incorrect connection the speakers are opposing one another.

14.8 Concert Hall Acoustics

Understanding how sound propagates in a room has an important application to the design of concert halls. Not only is the construction of a large concert hall very expensive, but if the hall is considered to have inferior acoustic properties, it is difficult to recover the large expense needed to engage first-rate performers. Thus, invariably the architect designs the hall in intense collaboration with physicists or engineers who act as their acoustic consultants. For economic reasons, concert halls need to be large enough to seat 2000 to 3000 people. During the last five decades, much progress has been made in the understanding of the design criteria that contribute to the quality of performance spaces for theater, opera, or orchestral music. Here we mention but a few of the most important considerations.

Outdoors, the sound from a solo violin or singer disperses in all directions so that already 100 feet away the sound is very feeble indeed, while in a good concert hall (e.g., Boston Symphony Hall, 150 feet long) the music is still heard clearly in the rear of the hall. Multiple reflections of sound from the walls of the hall are essential for sufficient loudness. Indoors, most of the sound that reaches the audience has undergone many reflections from the surfaces of the concert hall.

What do musicians and music lovers mean by *clarity, intimacy, envelopment,* and *warmth* of a concert hall? How are these desirable characteristics related to the architectural characteristics of the hall and aspects of the construction?

Clarity has an obvious meaning when we are concerned about understanding spoken or sung words, but clarity is equally important to discern each note of a rapid musical passage or to recognize separately the different instruments of an orchestra. Clarity requires sufficient volume of direct sound from the stage to the listener. Strong, direct sound is obtained with direct line of sight from the sound source to the listener. This is achieved using a raised stage and shallow balconies with raked seating—that is, seating that rises toward the back. The direct sound is often enhanced by structures hung over the performers and/or by panels placed behind the performers to direct the sound energy, which would otherwise be wasted, toward the audience. These reflectors take the form of canopies or are broken up into many smaller pieces referred to as clouds. The time taken by sound to travel to the clouds is small so that the sound reflected by the clouds can be considered part of the direct sound. To make the reflected sound disperse over the audience, the surface of the clouds tends to be convex rather than plane or concave. In addition, these structures help cohesion of the musicians in the orchestra. Figure 14.11 shows an example of clouds in a concert hall.

A concert outdoors or in an airplane hangar lacks intimacy. Performers have described an intimate hall—for instance, Carnegie Hall in New York, despite its 2804 seats—as one that "wraps its arms around you." The feeling of being in an intimate setting requires reflections from surfaces on the left and the right that follow the direct sound within a short time. Strong lateral reflections within 30 ms after the direct sound are characteristic of good acoustic design. Reflected energy beyond about 80 ms is sensed as a disturbing reflection or, for longer delays, as a distinct echo. The Symphony Hall in Boston is 24 m (70 feet) wide, so sound reflected from the side walls takes about $12/344\,s = 35$ ms to reach the center of the hall.

FIGURE 14.11 ▶ The donut-shaped clouds over the orchestra stage in the Opera House, Sydney, Australia.

After the early reflections, the sound in the concert hall should linger for the right amount of time. This decay of sound results from many reflections of sound from the surrounding walls, ceilings, seating, and so on, and arrives from all directions so that the listener feels enveloped in sound. The decay should be smooth, like that of a tuning fork (Chapter 12). The rate of decay is described by a number called the reverberation time (RT), which is the time required for sound level to decrease by 60 dB (see Chapter 15), which roughly corresponds in musicians' terms from very loud (fff) to very soft (ppp). To picture the meaning of the RT, imagine being in a still, large cathedral or a large vault with concrete walls, such as a large empty indoor swimming pool. The sound from a loud clap will linger for several seconds before it becomes inaudible, in contrast with a small bedroom, where the sound is swallowed up very quickly. In fact, the RT is measured, for example, by firing a toy pistol cap and recording the reverberant sound with a microphone and oscilloscope. You can try this yourself if you have a laptop computer with a sound card and suitable oscilloscope software, which is available free of charge from the web.

The desired reverberation time depends on the use of the hall and to some extent on musical taste. For speech, a shorter RT (less than 1 s) is desired for intelligibility than for classical music (1–2 s). Different composers (and conductors) favor somewhat different concert halls. Famous examples are Mozart and Beethoven, who preferred to conduct in the Wiener Hofburg with 1.4 sec reverberation time, while later composers of the Romantic period, such as Brahms, Bruckner, and Mahler, preferred the Musikvereinsaal Vienna with a rather long 2.1 sec reverberation time. The best reverberation time depends on the size of the hall (shorter for small halls) and on the sound frequency. A longer reverberation time for low frequencies (100 Hz) contributes to the sensation of warmth of a concert hall.

Of fundamental importance to acoustic design is the ability to calculate the reverberation time T_u by the Sabine equation. The important parameters of this equation:

$$T\,(in\ sec) = 0.161\,V/A\,(metric\ units),$$

or:

$$T\,(in\ sec) = 0.049\,V/A\,(U.S\ customary\ units)$$

are the volume of the hall V (in m^3 or in ft^3) and the total absorption area A (in m^2 or in ft^2) of the surfaces. If the walls absorbed no sound at all, $A = 0$, and the sound reverberates forever. In an actual room, different parts (walls, ceiling, seats, etc.) are more or less sound absorbing, so the absorbing area A is calculated as

$$A = S_1 a_1 + S_2 a_2 + \ldots$$

where the total surface of the hall is subdivided into various parts $S_1, S_2, \ldots \ldots$ and so on, with relative sound absorption coefficients $a_1, a_2, \ldots \ldots$. Wooden theater seats do not absorb much sound ($a = 0.07$ at

1 kHz), while the same seats occupied by people absorb about 10 times more ($a = 0.76$)! This is bad, because the reverberation time would change a lot with the room occupancy. Upholstered seats are much better because their absorption is high no matter whether they are occupied of not.

If a concert hall has too much absorption the sound is "dead." The problem is particularly severe if the hall lacks in warmth because of too much low-frequency absorption. Construction materials that particularly absorb low frequencies are avoided (e.g., plywood panels over empty spaces), while walls of concrete or plaster yield good bass reflection and warmth. Too little absorption is less of a problem because the architect can add sound-absorbing elements (cloth). Extensive tables of sound absorption coefficients for a large number of materials at different frequencies are readily available.

Not only must the sound in the hall last just the right length of time, but the sound must be distributed uniformly throughout the hall. Uniformity means there are no bad seats; everyone hears the music equally clearly. One essential tool to achieve uniformity as well as smooth decay of the reverberant sound is the use of diffusers. Large, flat walls tend to reflect sound in certain directions and can lead to sound bouncing back and forth from one surface to another (*flutter*). Diffuse reflection of sound requires irregular surfaces of different sizes, small for diffusion of high frequencies (short wavelengths), large for low frequency sound (large wavelengths). Thus, balconies with convex surfaces, sunken wall panels, ceilings with structural elements of different sizes, chandeliers, and so forth are desirable. Make it a habit when you wait for the concert to look around and identify as many sound-diffusing elements as you can.

Another virtue of a good concert hall is the absence of extraneous noise (street noise, ventilation, air conditioning). This sometimes requires large efforts. The Bridgewater Concert Hall in Manchester, England, is immediately adjacent to a railroad track. To isolate the building from the vibration caused by passing trains, the entire building is on a gigantic slab, supported on shock absorbers.

Until rather recently, the only way to ensure good acoustics was to copy the basic design of earlier, successful halls. The classic design is referred to as a "shoe box"—a rather large shoe box, of course, to which doors, a stage, seats, and balconies are added. But now it appears that computer-aided acoustic design has allowed successful deviations from the shoe box, such as the recently constructed Tokyo Opera City Concert Hall and the "Parco della Musica" Auditorium in Rome, Italy, which have been called acoustic miracles. Just what fraction of the recent successes can be credited to computations and how much to insights derived from long experience is still being debated.

14.9 Questions

1. You dip your finger in a still pond to make steady waves. If you wiggle the finger up and down 5 times every second, you observe wave crests that are 3 cm apart. What do you conclude about the speed of waves on the pond? Can you explain the answer to a friend who has common sense but knows no math or physics?

2. Are the waves propagating in a flute transverse, longitudinal, or both? What about the waves propagating on a string?

3. The range of human hearing extends from about 20 Hz to 20,000 Hz. Find the wavelength of the lowest frequency you can hear. Find the wavelength of the highest frequency you can hear. Find the frequency of a tone whose wavelength is 0.75 m.

4. How many times greater or smaller should the musician make the tension F on the string to double the speed of propagation of a wave?

5. The longest string on a concert grand piano is about 2 m long and has a mass of 50 g. (a) Find the mass per unit length of the string. (b) If the tension on the string is 300 N, what is the speed of wave propagation on this string?

6. At the Steinway piano factory, a worker finds that 500 m of piano wire on a spool weighs 2 kg. (a) Find the mass per unit length. (b) Find the speed of propagation on this wire for a tension of 40 N. (c) How much tension is required to double the speed of propagation?

7. Consider two loudspeakers oscillating in phase playing the same tune. Your ear is 6.0 m away from one speaker and 5.4 m from the other. Which frequencies will be suppressed? Which frequencies will be loud?

Chapter 15

Sound Perception: Pitch, Loudness, and Timbre

Now that we have learned about the basic properties of sound, let us begin studying what effect sound has on our ears. For the present purposes, it is more convenient to use electronic circuits for sound production rather than a musical instrument because we can more easily control and measure frequency, amplitude, and wave shape. Let us connect a loudspeaker (or headphones) to the signal generator (an electronic "oscillator") and display the wave shape on an oscilloscope.

It is easy to discover relationships between physical stimulus (what you observe on the oscilloscope) and the perception of the sound. Amplitude of the sound waves corresponds to the perceived *loudness*, while the frequency is perceived as *pitch*.

If one changes the shape of the wave from sinusoidal to triangular or square wave, the timbre is changed, but the pitch is not. *Timbre* (pronounce it in French, or as in the first two syllables of the word *tambourine*) is the "quality" of the sound. If you listen to a flute or violin both playing the same tone (same pitch) with the same loudness, you can immediately tell that they sound different. The aspect that distinguishes them is called the timbre or, by analogy with light, the physical "tone color." The timbre is associated with the *wave shape* of the sound pressure variations, and this differs between violin, flute, and any other string or wind instruments playing the same tone. We will later see that the wave shape of the tone is determined by its harmonic content. The way the tone rises in time (the attack) and the way the oscillations

Physics in the Arts, Revised Edition

diminish when the tone ends are also of much importance. These relationships can be shown schematically as

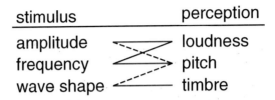

The black lines suggest that loudness depends not only on amplitude but also on frequency, and for the same frequency very loud and very soft tones are perceived by some people as having slightly different pitches. More detailed experiments on human hearing suggest that to a small extent all three aspects of the stimulus affect all three aspects of perception, but the red and solid black lines in the preceding diagram represent the most important aspects of tone perception.

15.1   Loudness and Amplitude

No doubt, when one increases the amplitude of the pressure variations by turning up the volume control on the radio, the perceived loudness increases. But does the sound get twice as loud when you make the amplitude twice as large? This question reveals a fundamental problem of perception studies: How do you know exactly what "twice as loud" really means, and would tests on different people who are asked to judge "twice as loud" in an experiment give similar results?

A much more objective test on loudness perception is done by playing two tones of the same frequency (e.g., 1000 Hz) but slightly different amplitude one after the other. The difference in amplitude is then made smaller and smaller until the listener can no longer tell the difference. This is done by alternating between the softer (S) and the louder (L) tone randomly (L L S L S S S L S etc.) and the listener indicates which tones sound louder or softer. In this way one finds the JND (just noticeable difference) in loudness perception.

The results on loudness perception are surprising in two respects:

1. We are rather insensitive to small differences in amplitude. Increasing the amplitude 1.5-fold is only barely noticeable.

2. A given increment in loudness corresponds to a certain **ratio** of sound pressure amplitude. For instance, a change in amplitude from 10 to 15 units sounds the same to us as the change from 20 to 30 units, or from 100 to 150 units. Note that from 10 to 15 the amplitude changed

by 5 units, and from 100 to 150 the amplitude changed by 50 units, but subjectively the loudness change is the same in both cases. **Equal increments in loudness** correspond to **multiplying the amplitude by the same number**. Many audio systems let you increase the volume in steps. These are equal steps in loudness, each time multiplying the amplification by the same factor.

Another surprising aspect of our loudness perception is the vast range of amplitudes that we can perceive, from the softest tone to the loudest tone we can tolerate. Sound pressure amplitude is measured in N/m^2—it tells how much force (in Newtons) is pushing against the eardrum per unit area (pressure = force/area). For a frequency of 1000 Hz, for the average person of normal hearing, the threshold of hearing is a sound pressure amplitude of $0.00003 \ N/m^2 = 3 \times 10^{-5} \ N/m^2$. Compare this pressure to that exerted by a single sheet of paper on the table: the threshold of hearing is 20,000 times less pressure! But at the other extreme, the loudest sound you can still stand, although painful, exerts a pressure 1,000,000 times greater than the threshold of hearing—that is, a pressure of 30 N/m^2 (50 sheets of paper).

Sound intensity means how much sound power hits one square meter of area that absorbs the sound. Power is measured in watts (abbreviated W, such as in a 60 W lightbulb). *Sound intensity (in W/m^2) is proportional to the square of the sound pressure amplitude*. Without going into details, a similar relationship holds between the amplitude and energy of a pendulum or a mass oscillating on a spring: Doubling the amplitude quadruples the energy because for twice the amplitude the spring gets stretched twice as far, but in addition it is twice as hard to pull: $2 \times 2 = 4$—the square.

How large is the sound intensity for the loudest sound you can tolerate, compared to the threshold of hearing? The sound pressure amplitude is 1,000,000 times larger, so the **sound intensity** is the square of that—**1,000,000,000,000** times larger, or 10^{12}!

Table 15.1 shows some of the important aspects of what we just talked about. The table lists different typical intensities of sound sources, from a jet engine to the softest sound you can barely hear in a soundproof enclosure (the threshold of hearing). In each case the sound pressure amplitude, the ratio of the sound pressure to the threshold sound pressure (p_o) for a 1000 Hz tone, and the corresponding intensity level I in W/m^2 are listed. The column labeled I/I_o gives the sound intensity relative to threshold intensity I_o.

For the loudest sound, the intensity is, as mentioned, 10^{12} times larger than the threshold intensity.

Remember that equal increments in subjective loudness correspond to equal **factors** of sound amplitude or sound intensity. For each step in Table 15.1, the sound pressure increases 10-fold, and the intensity, being the square of the sound pressure, increases 100-fold.

TABLE 15.1 ▶ Intensities of Different Sound Sources

Description and Musical Notation	Pressure Amplitude p in N/m^2	p/p_o Pressure Relative to Threshold	Intensity I in W/m^2	I/I_o	Sound Level in dB
Jet Engine at 20 feet Loud as Hell	30 3×10^1	1,000,000 10^6	2	10^{12}	120
Rock Band Fortissimo (fff)	3 3×10^0	100,000 10^5	0.02 2×10^{-2}	10^{10}	100
Very Heavy Traffic Loud f (forte)	0.3 3×10^{-1}	10,000 10^4	0.0002 2×10^{-4}	10^8	80
Quiet Conversation Soft (p—piano)	0.03 3×10^{-2}	1,000 10^3	0.000002 2×10^{-6}	10^6	60
Quiet Room Very Soft (ppp)	0.003 3×10^{-3}	100 10^2	2×10^{-8}	10^4	40
Whisper at 4 feet Hard to Hear	0.0003 3×10^{-4}	10 10^1	2×10^{-10}	10^2	20
Threshold of Hearing at 1000 Hz	0.00003 3×10^{-5}	1	2×10^{-12}	1	0

It seems that a better measure of loudness is, therefore, the number of zeroes in the intensity level relative to threshold. The intensity of the loudest sound (relative to threshold) has 12 zeroes, while the softest (threshold) has none. This indeed is how the physical sound intensity level is specified, except that the number of zeroes is multiplied by 10. **This is, by definition, the sound level in decibels (abbreviated dB):**

Sound level in dB $\boxed{\mathrm{dB} = 10 \log \frac{I}{I_o}}$.

$10 \log(I/I_o)$ is a fancy mathematical way of indicating that we take the number of zeroes in the I/I_o ratio. Calculated this way, then, the loudness in the last column of the table shows that the loudest tone has 120 dB, the threshold is 0 dB, and in between equal increments in the number of dB correspond to equal increments in subjective loudness.

One way you use this equation is to put the number for I/I_o in your calculator and then press the "LOG" button. Try it; if you put in 100, you should get the answer 2, because there are two zeroes. To get the decibel level, multiply this answer by 10.

We will be using logarithmic scales again later when we talk about musical scales. It may be useful to show how numbers on a logarithmic axis are distributed:

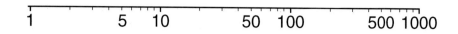

In contrast to the normal number line, where equal increments in the number are represented as equal distances, here **equal factors** are represented by **equal distances**. Note, for example, that a 10-fold increase (from 1 to 10, or from 5 to 50, or from 10 to 100) shows up as the same length. Perceived loudness increases the same amount for every 10-fold increase in sound intensity level. Thus, you can think of perceived loudness represented by the distance along a log number line, while the numbers along the line correspond to the intensity level.

We mentioned earlier that we are not very sensitive to small changes in sound intensity level. Under good laboratory conditions, a 1.5 dB change is just noticeable, but in real life a 3 dB change is needed before you can say a tone is clearly louder. To calculate I/I_o for 3 dB, you put 3 into the calculator, divide by 10, and press the inverse function of log, called 10^x. The result you get is 2.0. This tells you that doubling the sound intensity level is needed to bring an appreciable increase in subjective loudness.

If you have a good amplifier, the volume control may be labeled "loudness" and the numbers given on the dial or the digital display are the dB numbers. Sound intensity levels in dB are always with respect to some arbitrary reference. In the previous equation, the reference level is the threshold of hearing, but dB is a useful number to compare *any* two intensity levels. On an amplifier, the reference level is rather arbitrary, often chosen for the very highest amplification to be called 0 dB and to specify lower amplification by negative numbers.

15.2   Loudness and Frequency

A simple experiment demonstrates that perceived loudness depends not only on amplitude but on frequency also. If one slowly changes the frequency of a tone with fixed amplitude from 1,000 to 30,000 Hz, the perceived loudness of the sound decreases sharply in the neighborhood of 15,000 Hz. Above 22,000 Hz one hears nothing, although a microphone and oscilloscope would confirm that there is a sound wave present. The air is transmitting pressure waves, but the frequency is too high for our ears to hear; it is "ultrasound." With increasing age, one hears less and less of the high frequencies; as early as age 25, the ability to hear very high frequencies declines. At low frequencies, hearing extends to about 20–30 Hz; for large amplitude sounds below 20 Hz, you may still perceive some sort of pressure variation but not a tone of a definite pitch.

If we compare sounds with different frequencies, how is their perceived loudness related to the intensity, or the amplitude, of the sound waves? By switching back and forth between two tones of different frequencies and adjusting the amplitude until they seem about **equally loud**, we can study how loudness perception changes with frequency. Do two tones at different frequencies sound equally loud when they have the same amplitude—that is, the same intensity level I in W/m^2?

The answer is, not at all!

The results of such measurements are called the Equal-Loudness curves. To explain the method, we show one example in Figure 15.1. One uses as a reference a pure tone of 1000 Hz and adjusts the sound intensity level to the desired value—in this example 40 dB. The perceived loudness of this tone is called loudness level 40 *phons*. Tones at other frequencies that are judged to sound "equally loud" are also said to have a loudness level of 40 phons. What the measurements show is that the physical sound pressure or sound

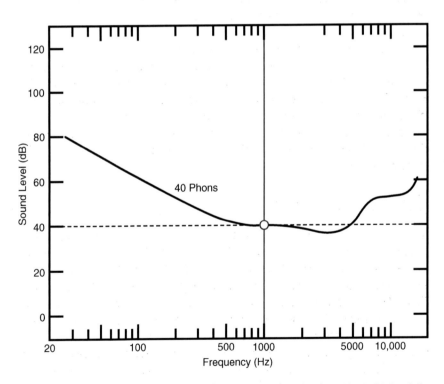

FIGURE 15.1 ▶ Curve of equal loudness: the curve shows the sound intensity level (in dB) that is needed to match the loudness of a 40 dB tone at 1000 Hz. All the tones along the curve have a loudness level of 40 phons.

intensity level is very much different for different frequencies—that is, the ear has a different sensitivity at different frequencies. Figure 15.1 shows that, for instance, at 100 Hz, the sound intensity must be 60 dB to sound as loud as a 1000 Hz tone of 40 dB intensity. This difference of 20 dB is very large if you remember that 20 dB means 100 times greater intensity level (or 10 times larger sound pressure level). The ear is *much* less sensitive at 100 Hz than at 1000 Hz.

An interesting result is found when the same experiment is repeated starting with varying loud sounds at 1000 Hz. Figure 15.2 shows an entire set of measurements of perceived loudness as a function of frequency for different intensities 10 dB apart at 1000 Hz. When the tone is very loud, such as 90 dB, the curve is almost flat, indicating that for very loud sounds the sensitivity of the ear does not depend much on frequency. For loud sounds the ear is equally sensitive to different pitches. But at the other extreme, for very soft

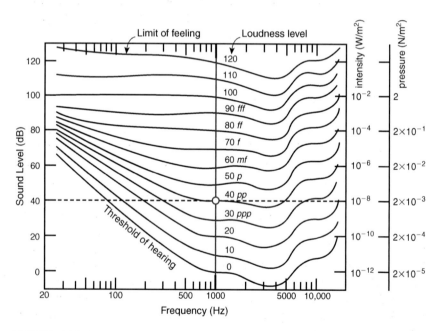

FIGURE 15.2 ▶ Tones at 1000 Hz in 10 dB increments of sound intensity level are matched in loudness by tones of frequencies between 20 Hz and 20,000 Hz. The corresponding loudness levels of musical tones are shown in the usual notation (*f* stands for forte, etc.). Sound intensity level (dB) is on the left axis, and intensity (W/m²) and the corresponding pressure amplitude (N/m²) are on the right. The graph is adapted and reused with permission from H. Fletcher and W. A. Munson, *The Journal of the Acoustical Society of America*, **5**, 82 (1933). Copyright © 1933, Acoustical Society of America.

sounds, there is a tremendous difference in sensitivity: The sound needs to have more than a 1000 times greater intensity to be barely audible at 100 Hz than at 1000 Hz. Figure 15.2 shows that hearing of faint sounds is most acute in the range of about 1000 Hz to 6000 Hz.

You might wonder if any of this is of practical use. One application that will interest you is in music reproduction. Suppose you like to listen to music played at an ear-splitting loudness of 90 dB. Now your neighbors complain, so you turn down the volume on the amplifier. While for loud sounds the ear was nearly equally sensitive for different frequencies, after you turn down the amplifier from 90 dB to 50 dB, the midrange is still audible, but the lower frequencies have almost completely disappeared, because at low sound intensity levels low frequencies are harder to hear. So you are very unhappy about the sound, unless you turn up the bass quite a lot. Most of the better amplifiers do this automatically; that's why the knob is labeled *loudness* control and not volume control.

A similar comment applies to the conductor of a band or an orchestra. Since the conductor is close to the orchestra, his sound intensity level is much greater than for the audience, so he needs to compensate by "turning up" the volume of the bass and of the high frequencies.

15.3 Pitch Discrimination

For musicians it is important to know how small a difference in frequency a person can recognize. We learned in Chapter 13 that beats between two tones allow you to detect even the smallest frequency differences when the two tones are played at the same time. Here we ask how well a person can tell when two frequencies are slightly different, one tone played *after* the other.

The answer depends on many factors. Very short tones can be off more than long, steady tones, which is a good thing for the violinist because if the pitch is a little off-key in a very rapid passage, one is less likely to notice than for steady tones. For test tones of one-second duration, experienced musicians can tell when the frequency between two successive tones differs by about 0.3% or 3 Hz for a 1000 Hz tone. So you see that the ear is very good at detecting differences in frequency between two tones, but not good at recognizing differences in sound intensity.

Chapter 16

The Ear

The eye and the ear are probably the most complicated sensory instruments in our body—as well as the most remarkable for their ability to perceive and process a vast range of stimuli. Here we give a brief, simplified review of the way the ear senses sound pressure variations and converts them into nerve pulses transmitted to the brain. Does the ear work like a microphone, which accurately changes pressure variations into an electric signal of the corresponding wave form? How does it manage to work over such a vast range of intensities? What limits the frequency range of the ear? How is it possible for the ear to distinguish tones which differ in frequency by less than 1%? These are but a few of the questions one might ask when one thinks about the ear as a sound detector. In passing we might mention that the ear not only allows us to hear sound, but it is also the organ that helps us maintain balance.

16.1 The Parts of the Ear

The components of the ear, shown in Figure 16.1, can be grouped into three parts: *the external ear, the middle ear, and the inner ear.*

The external ear consists of the pinna and the ear canal. The pinna is the outer part of the ear, while the auditory canal is the tubular opening leading to the middle ear. Although the pinna is not as important to us as it is to many other mammals (elephants!), it helps us distinguish the direction from which the sound originates. The pinna alters the intensity of sound from a source behind our head compared to that from a source in front of us. In addition, the pinna collects sound and guides the sound waves more or less toward the ear

Physics in the Arts, Revised Edition

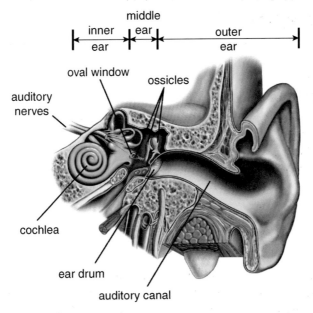

FIGURE 16.1 ► The parts of the ear. Medical Illustration Copyright © Nucleus Medical Art, All rights reserved. *www.nucleusinc.com.*

canal—more or less, because you remember that in considerations of sound propagation (Chapter 14), one should compare the size of the object to the wavelength of the sound. The pinna is approximately 6 cm in size. The frequency that corresponds to a 6 cm wavelength is about 6,000 Hz, so the pinna helps sound collection primarily for the higher frequencies.

The ability to tell from which direction sound originates has two aspects: the difference in sound volume received by the left and the right ear, and the tiny difference in the arrival time of the sound between the two ears.

At the end of the auditory canal begins the middle ear, which consists of the tympanic membrane (eardrum) and three tiny bones (ossicles) transmitting the motion of the ear drum to the inner ear. These bones are called the hammer, the anvil, and the stirrup, according to the objects their shapes resemble. These are the smallest bones in the human body.

Together, the ossicles act like a lever or a crowbar: The eardrum pushes back and forth at one end (hammer), and at the other end (stirrup), the amplitude of motion is less but the force is greater. What is quite remarkable is that our body has the ability to change the effective fulcrum (point of rotation) of the lever, so for very loud sounds the ear is partially protected from damage. This involves a feedback mechanism in which the brain, when a sound is too loud, makes muscles in the ear change the lever action!

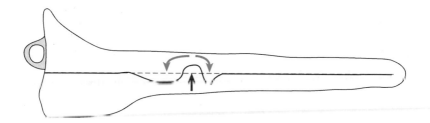

FIGURE 16.2 ▶ Longitudinal section through the stretched-out cochlear duct. The entire cochlea is approximately 35 mm (or 1.4") long. Pressure from the eardrum is transmitted to the stirrup (top left), which presses on the oval window. The pressure wave travels along the top of the basilar membrane through the fluid inside the cochlea and returns along the bottom of the membrane back to the round window on the lower left. In the process, the basilar membrane is deflected, but note that the amplitude of the membrane deformation is much exaggerated in this schematic.

The inner ear contains the cochlea, which converts the mechanical pressure variations into nerve signals. The cochlea is a spiraling, tapered, snail-like, bony structure. It is filled with fluid and has two windows. The "oval window" (Figure 16.1) is attached to the stirrup, which wiggles the window back and forth and sets up a pressure wave in the liquid inside the cochlea.

The function of the cochlea is much easier to describe if we imagine the snail-like spiral stretched out, as shown schematically in Figure 16.2. The basilar membrane is a flexible membrane that runs down the middle of the cochlea and divides it into an upper and a lower part, connected by an opening at the apex of the cochlea.

16.2   Place Theory of Pitch Perception

Attached to the basilar membrane are about 15,000 nerve cells, which respond to the bending of the membrane. The membrane is much more flexible near the apex than near the oval window. Depending on the frequency of sound, the flexing of the membrane shifts from one region of the cochlea to another, as shown in Figure 16.3. Correspondingly, different nerve cells are stimulated for sounds of different pitches. For example, a 1600 Hz tone mostly stimulates cells between 10 and 20 mm from the oval window but almost no cells beyond 22 mm. On the other hand, low-frequency tones stimulate nerve cells near the flexible, floppy end of the basilar membrane.

The *place theory* of sound perception holds that we recognize the pitch of a tone by the location on the basilar membrane of the groups of nerve cells that are affected by the sound. While this plays a role, place theory would be

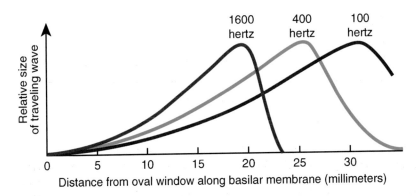

FIGURE 16.3 ▶ For low-frequency tones, the basilar membrane flexes mostly near the apex for high-frequency sound nearer to the oval window. Correspondingly, different nerve cells are stimulated for sounds of different pitches.

hard pressed to explain why almost anyone can tell apart tones that differ by 1% in frequency, and musically trained people do even better than that. If one imagines shifting the 400 Hz curve by 4 Hz (1% of 400), the shift is so small that it can hardly be the principal way we recognize pitch. Recent studies of signals from single nerve cells provide a better explanation.

16.3 What Do the Auditory Nerves Tell the Brain?

The nerve cells in the cochlea send signals to the brain, where the signals are processed into what we call *hearing sound*. While the processing in the brain is not well understood, fascinating research on the nerve response of single auditory nerve cells has been done at the University of Wisconsin Department of Neurophysiology.

In general, nerve cells are known to communicate with the brain by sending electrical signals not in the form of a complex wave shape but by sending a sequence of individual nerve impulses.

Studies on the response of individual auditory nerve cells have found remarkable results, briefly summarized here:

1. When the auditory nerve is stimulated by a periodic signal, such as a pure tone (sine wave), the nerve pulses occur at various, **seemingly irregular** intervals. All nerve pulses ("spikes") are nearly of the same height, independent of the loudness of the tone.

2. A histogram of the spacing between adjacent pulses shows an interesting **regularity:** The pulses are grouped at multiples of the period T of the periodic tone (Figure 16.4).

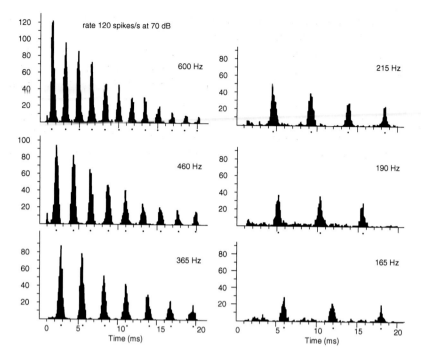

FIGURE 16.4 ▶ The histogram shows the distribution in spacing between adjacent nerve impulses from a single auditory nerve cell stimulated by sinusoidal tones of different frequencies. The peaks in the distribution are at multiples of the period of the sound. The histogram for 600 Hz shows that sometimes 10 or more periods pass before the next nerve impulse occurs. The average rate of the nerve impulses (spikes) is related to sound intensity level (dB). The figure is adapted and used with permission from a paper by J. E. Rose et al. in the *Journal of Neurophysiology,* **30**, 769 (1967).

3. The loudness of the tone changes the **average** rate with which the pulses occur. The rate is related to the sound level in dB of the tone.

4. Nerve pulses are transmitted only during the positive part of the pressure wave, not during the wave troughs.

5. The wave shape is encoded in the detailed shape of the nerve pulse peaks in the histogram in Figure 16.4. For a pure tone, the peaks of the distribution reflect the positive part of the sine wave. For a complex tone, the distribution changes correspondingly.

When listening to different sounds, the brain receives a sequence of nerve pulses (spikes) with different time distributions from about 15,000 nerve cells in the cochlea. Since we know that we can distinguish changes in pitch of 1% or less, we conclude that the brain must analyze all these histograms and determine the periodicity of the spike distribution to high accuracy. Therefore, for very soft tones, when the spikes in the average are few and far between, it must take longer to accumulate sufficient information to recognize the pitch accurately. This is indeed the case: the accuracy of pitch perception for very short tones of very low levels is much poorer than for louder and longer tones.

This brief summary shows that the auditory system encodes loudness, pitch, and timbre in a complicated way. We don't fully know how the brain turns this stream of encoded nerve pulses into sensations. You might, however, stop and marvel at your ear and brain functioning the next time you hear a sound.

Chapter 17

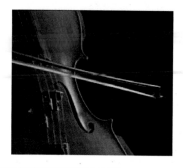

Vibration of Strings

17.1 Single Modes

The simplest kind of oscillation of a guitar string is a harmonic to and fro motion, just like the one shown in Figure 17.1 for a Slinky or for a string in Figure 17.2. In this case, each point of the Slinky or the string moves up and down in simple harmonic motion, and all points oscillate with the same frequency but with different amplitudes. When all parts of an oscillating system vibrate in simple harmonic motion, the motion is called a Mode of the oscillating system. Whether the oscillating system is a string or a drum, all of its parts vibrate with *the same frequency*. As we will see later, the actual motion when a string is plucked or bowed is more complicated, but we will be able to understand these complicated motions as the simultaneous occurrence of several modes.

Only certain special frequencies are allowed. To set up such an oscillation, you would gently wiggle the Slinky or the string up and down in sinusoidal motion. The wave you create in this way travels along the string and is reflected at the far end where the string is tied down. The wave then travels back toward you and is reflected once more at your end of the string, and on and on, the wave goes back and forth. Notice one important point that is essential to understand why only certain frequencies are possible on a string: To build up a steady oscillation, each subsequent wave crest that you make with your hand should be in step with the wave already present on the string. This means that you should time the wave crests to be spaced exactly equal

Physics in the Arts, Revised Edition

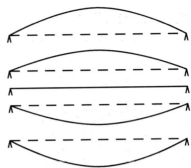

FIGURE 17.1 ▶ Top: The shape of a Slinky at two instances in time when it oscillates in the first (fundamental) mode. Bottom: Photograph of a string in vibration.

FIGURE 17.2 ▶ The shape of a string at different instances in time when the string vibrates in the first (fundamental) mode. The four drawings are one-eighth of a period ($T/8$) apart in time and represent half a period ($T/2$).

to the round-trip time of the wave on the string. A whole number of complete periods (1, 2, 3, etc.) must fit into one round-trip time. Conclusion: *in the fundamental mode of oscillation of the string, the period T equals the round-trip travel time of the wave on the string*. If the length of the string is called L, and the speed of propagation of the wave on the string is v, then the period of oscillation T (round-trip travel time) is $T = 2L/v$. Since the frequency is the reciprocal of the period T, or $f = 1/T$, then

---

First (fundamental) mode frequency: $\quad f_1 = v/2L$

The equation says that *shorter string corresponds to higher frequency* (string length is inversely proportional to string oscillation frequency). When the musician shortens the string by placing a finger on the fingerboard the pitch increases. Note that the speed of propagation v on the string in the formula stays the same as long as the musician does not change the tension (pull on the string) with the tuning pegs.

You can do this yourself. Find the round-trip travel time of a pulse on a Slinky stretched between you and a friend. Then move your hand gently up and down in step with the reflected pulse and observe that you build up a mode that looks like Figure 17.1.

Example: The string of a guitar is 60 cm long and is tuned to standard A = 440 Hz. What is the new frequency if the vibrating length of the string is shortened to 40 cm by placing a finger on the fingerboard?

Answer: Frequencies are inversely proportional to the string lengths. The ratio of lengths is 40 cm/60 cm = 2/3. Thus, the new frequency is (3/2) times the old frequency = (3/2) × 440 Hz = 660 Hz. The new frequency has to be higher, since the string is shorter.

17.2   Higher Modes

What happens if you gently wiggle your hand exactly twice as fast? If you understood the argument in the preceding section, you will anticipate the answer: You now make not one but two full waves for every round-trip distance of the string so that you are again reinforcing the existing oscillation. So the next higher allowed frequency is twice the fundamental frequency, or $f_2 = 2f_1$ (see Figures 17.3 and 17.4).

   If you shake the Slinky (or the string) even faster, you can produce three half-wavelengths along it. Again all points of the Slinky make simple harmonic motion (see Figures 17.5 and 17.6).

   The third mode has frequency $f_3 = 3f_1$.

   The n-th mode has a frequency n-times the fundamental frequency, or

$$\boxed{f_n = nf_1 = nv/2L}$$

Wavelength: If you look at the picture of the string oscillating in the first mode (Figure 17.1), it looks like half of a wavelength fits into the length of the string. So we can also express the condition for the fundamental oscillation by saying that $\lambda_1/2 = L$.

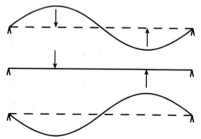

FIGURE 17.3 ▶ Top: Oscillation of a Slinky in the second mode. Bottom: Photograph of a string oscillating in the second mode.

FIGURE 17.4 ▶ Shape of the string at different instances in time. The pictures are 1/4 period apart.

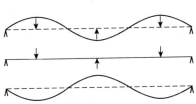

FIGURE 17.5 ► Top: Oscillation of a Slinky in the third mode. Bottom: Photograph of a string oscillating in the third mode.

FIGURE 17.6 ► Shape of a string at different instances in time. The drawings are 1/4 period apart.

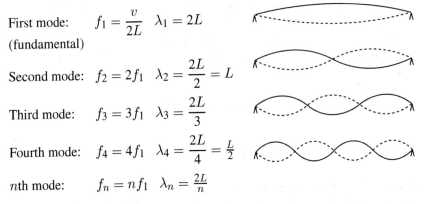

First mode: $f_1 = \dfrac{v}{2L}$ $\lambda_1 = 2L$
(fundamental)

Second mode: $f_2 = 2f_1$ $\lambda_2 = \dfrac{2L}{2} = L$

Third mode: $f_3 = 3f_1$ $\lambda_3 = \dfrac{2L}{3}$

Fourth mode: $f_4 = 4f_1$ $\lambda_4 = \dfrac{2L}{4} = \dfrac{L}{2}$

nth mode: $f_n = nf_1$ $\lambda_n = \dfrac{2L}{n}$

FIGURE 17.7 ► The oscillation frequency f and the wavelength on the string when the string is oscillating in one of its modes.

Well, yes, this is correct: Remember that $\lambda = v/f$, so if we substitute for f, the frequency of the first mode ($f_1 = v/2L$), we get $\lambda = 2L$.

For the n-th mode, the frequency is n times higher so that the wavelength is n times shorter. Thus, n half-wavelengths fit into the string length.

This is summarized in Figure 17.7. For all modes, each point of the string makes simple harmonic motion, but the frequency is now twice, three times, four times as large as for the first mode.

17.3 Traveling Versus Standing Waves

We should realize that these motions of the string in the various modes are quite different from the waves we have encountered before. In a *traveling*

FIGURE 17.8 ► In a traveling wave, the wave crests move along at speed v. All points move in simple harmonic motion with the same amplitude and the same frequency but shifted in phase.

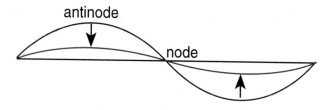

FIGURE 17.9 ► In a standing wave, all points between adjacent nodes oscillate with the same frequency, but, in contrast to the traveling wave, all points oscillate in phase but with different amplitude. Points that never move are called ***nodes***, and points that have the largest amplitude are called ***antinodes*** (or "bellies").

wave the wave crests moved along the water's surface; all points of the system (the water surface) moved up and down, and nothing stood still. This is represented in Figure 17.8.

In contrast, the wave pattern of the Slinky or the string is a *standing wave;* it moves neither to the left nor to the right but is the sum of two waves moving in opposite directions (see Figure 17.9). Here are the differences between traveling waves on a long string and standing waves on a string tied at both ends:

- In a traveling wave the crests and troughs move along the string.

- In a traveling wave all the points move up and down with the same amplitude. Different points have different phases.

- In a standing wave the amplitude is different at different points. Some points don't move at all. These places are called ***nodes*** of motion.

- In a standing wave points of maximum motion (bellies or ***antinodes***) are halfway between the nodes. All points between two nodes move ***in phase***.

17.4 The Voicing Formula

To find the frequencies of the string's modes, we must know the speed of propagation v of the wave on the string. But we have learned that v depends on the tension and the mass per unit length of the string (Chapter 14). Using this result, we can find the fundamental frequency of a taught string, which is tensioned by a pull F

$$\text{Fundamental frequency:} \quad f_1 = \frac{v}{2L} = \frac{1}{2L}\sqrt{\frac{F}{M/\ell}} = \frac{1}{2L}\sqrt{\frac{F}{\rho}}$$

Remember that $\rho = M/\ell$ simply means the mass of a 1-meter length of string, no matter how long the actual string is. Piano tuners and piano builders call this the *voicing formula* because "voicing" a piano means to put strings on it; without strings the piano has no voice!

$$\text{Frequency of higher modes:} \quad f_n = \frac{n}{2L}\sqrt{\frac{F}{\rho}} = nf_1$$

The equations say that the frequency increases as the tension is increased, in proportion to the square root of the tension F. This means that to double the frequency, the tension must be 4 times larger. The formula says that the frequency is lower for a thicker string (larger M/ℓ), in inverse proportion to $\sqrt{(M/\ell)}$.

How does the piano builder choose the length, the tension, and the mass of the string? The tension should be as high as possible (without breaking the string!) so that the oscillating string presses hard against the soundboard, producing a good, forceful sound volume. Bass strings need to be long and thick (massive), since L and M/ℓ are in the denominator. In a large grand piano, the longest strings are about 7 feet long (2.1 m). But to make the frequency low enough, the strings still need to be quite massive. One must clearly distinguish between the wavelength of the wave propagating along the string, which depends on the speed of propagation on the string, and the wavelength of the piano sound in air, which depends on the speed of sound in air.

Example: A piano string is 1.2 m long and has a tension of 200 N. The tuner cut the string from a spool of wire that was labeled "200 m piano wire/mass 1.2 kg."

What is the fundamental frequency of this string and what are the frequencies of the next two higher modes?

Answer: We substitute $M/\ell = 1.2\,\text{kg}/200\,\text{m} = 0.006\,\text{kg/m}$; $F = 200\,\text{N}$, $L = 1.2\,\text{m}$

$$f_1 = \frac{1}{2 \times 1.2} \sqrt{\frac{200}{0.006}} = 76.1\,\text{Hz}$$
$$f_2 = 2 \times 76.1\,\text{Hz} = 152\,\text{Hz}$$
$$f_3 = 3 \times 76.1\,\text{Hz} = 228\,\text{Hz}$$

17.5   How Do Modes Relate to Music?

When a string is bowed or plucked, many modes of oscillation occur at the same time—that is, the actual motion of a bowed or plucked string is the *superposition of many modes*, often 20 or more modes.

Correspondingly, the tone produced by a 440 Hz string has in it also vibration frequencies of two times, three times, four times 440 Hz. Musicians call these the *higher partials* of the tone, or *overtones*. The fundamental mode is the first partial, the second mode is the second partial, and so on. Since *overtone* refers to the oscillations higher than the fundamental, the first overtone is the second mode (second partial), and so on.

The frequencies of the partials of a musical instrument need not be exact multiples of the first mode; in fact, for drums this is by no means the case, nor is it true for stiff vibrating bars (xylophone). Partials whose frequencies are exact multiples of the fundamental are called *harmonics*.

The frequencies of the harmonics form the harmonic series:

harmonic series: f_1 $2f_1$ $3f_1$ $4f_1$ $5f_1$ and so on

Note that the partials have simple frequency ratios: the second partial has twice the frequency of the first. In musical terms it is the *octave* (frequency *ratio 2:1*); the fourth partial is the octave of the second partial, and so on. The third and second partial form a *3:2 ratio*, which in music is called a *fifth*, as mentioned in Chapter 13. We will discuss this at length in Chapter 20, where we talk about musical intervals and musical scales.

As we just mentioned, the motion of a bowed or plucked string is complicated, but it can be thought of as a superposition of many modes. How much of each mode is present (its amplitude) depends on how and where the string is plucked or bowed. It may be interesting to see the motion of a string that vibrates in several modes at the same time. This can be done with a computer simulation; it may not work on all computers, but try by going to *http://surendranath.tripod.com/Applets/Waves/Harmonics/HarmonicsApplet.html*. The mix of partials determines the tone color (timbre) of the musical tone. The relative strength of the various partials can be represented by a bar graph, which shows the amplitude of each partial, as shown in Figure 17.10.

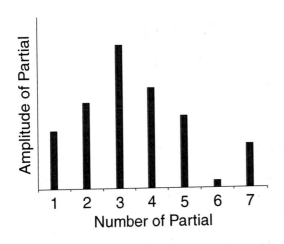

FIGURE 17.10 ▼
The spectrum of partials in a tone represented as a bar graph. The mix of partials determines the timbre of the musical tone. On the horizontal axis we indicate the number of the partial, while the height of the bar indicates its amplitude. A higher bar indicates a stronger partial (e.g., the third partial in this graph).

17.6 Damping of Higher Partials

After plucking a string, the sound slowly decays—that is, the oscillation is damped because energy is dissipated not only as sound but also in the back-and-forth bending of the string. If you listen carefully to the decaying tone of a guitar string, you may notice that the tone not only becomes softer (lower amplitude) but more and more the tone approaches a pure tone. This change in timbre is related to the fact that the *higher modes are damped faster* because they require more bending of the string.

17.7 Plucked Strings: Missing Partials

While it is not always easy to anticipate how much of each partial is present in the tone of the instrument, you may expect that to produce a strong first mode you have to pluck the string at its center because this is where the first mode has an antinode (belly).

If you pluck the string at the exact center, is the second mode present? No, because the second mode has a node at the center and is therefore incompatible with plucking at the center. Moving the string at the center does not excite this mode. Similarly, all higher modes with a node at the center are absent. The second, fourth, sixth, and eighth modes are all missing, as shown in Figures 17.11 and 17.12.

Generally speaking, modes that have a node at the plucking point will *not* be excited. Modes that have an antinode at the plucking point will be strongly excited. The strength of excitation of modes depends on whether the plucking point is close to a node (weak excitation = small amplitude) or close to an antinode (strong excitation = large amplitude). It is for this reason that a guitar

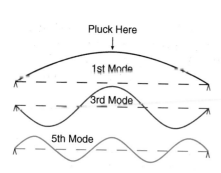

FIGURE 17.11 ▶ String plucked at the center: all modes with a node at the center are absent.

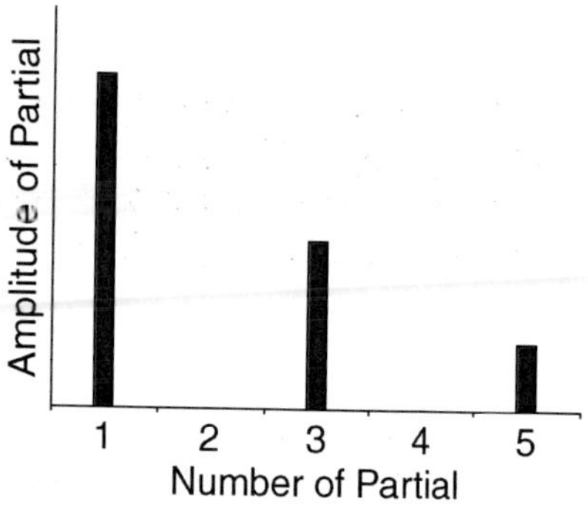

FIGURE 17.12 ▶ Spectrum of partials for a string plucked at the center shown as a bar graph.

string produces different sounds when plucked at different places; the closer to the bridge the strings are plucked, the more higher modes will be present, and the brighter the sound.

Try to figure out which modes are missing and which ones are strongly excited if you pluck the string at 1/3 of the string length (for instance, you pluck a 60 cm long string at a distance 20 cm from either end). From the distribution of partials in Figure 17.10, try to determine where the string was most likely plucked.

17.8 Playing Harmonics

When a string vibrates in only one mode, nodes and antinodes are located in a regular pattern along its length. If you touch the string lightly at one of the nodes, the vibration will continue undisturbed. If you touch it anywhere else, the vibration will stop instantly.

If one plucks a guitar string near the bridge, many modes are present in the sound. If the guitarist now lightly touches the string at a point 1/4 length from the end, which modes will still sound? Touching the string at a given point kills all modes except the ones that have a node where the finger is placed. The modes with a node where the finger was placed survive because the string did not move at that place anyway. Figure 17.13 shows which partials survive in this case.

17.9 Real Strings Have Some Stiffness

In this chapter we assumed that the speed of propagation of waves on the string, v, does not depend on the wavelength (or the frequency) of the waves.

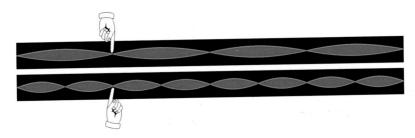

FIGURE 17.13 ► Playing harmonics: we assume that the string was originally plucked near one end of the string and that a finger is lightly placed on the string 1/4 length from the end. The picture shows that the fourth, eighth, and so on partials still oscillate.

This is true only if the string has no stiffness at all and all the restoring force is provided by the tension of the string.

Real strings, however, have some stiffness (gut strings very little, steel strings quite a bit) and this will be felt as an increase in restoring force for the higher modes (where there is more bending of the string). The stiffness of the string increases the velocity of propagation for the higher modes and therefore slightly increases their frequency. Musicians say that the higher modes are sharp (i.e., slightly too high in frequency).

We have learned that bass strings need to be massive—that is, they have relatively larger diameters, which makes them stiffer. How can one make a massive metal string that is still quite pliable—in other words, has a small stiffness? One uses overwound strings: A second wire is wound around a central core, thereby increasing the mass without increasing the stiffness. All bass strings on the piano and the lower strings on most string instruments are overwound, but of course some stiffness remains.

Typically, for a piano string, the second mode is sharp by about 0.2%; that is, if the fundamental is 100 Hz, the second partial is not 200 Hz but 200.4 Hz, the third partial is 300.6 Hz instead of 300 Hz, and so on. We say that the partials of the piano string are slightly *anharmonic*.

17.10 Questions

1. A guitar string of 60 cm length has a fundamental frequency of 240 Hz. One can increase the frequency to 300 Hz by placing a finger on the fingerboard so that the vibrating length becomes shorter. How long should the vibrating length be? What will be the frequency of the string if the vibrating length is 40 cm? 20 cm?

2. What happens to the frequency of a 200 Hz piano string if you changed the tension from 200 N to 50 N? From 200 N to 300 N? What happens

to the frequency of a piano string if you double the diameter? (You need to ask yourself, does the mass of the string double? Triple? Quadruple?)

3. If the tension of a 240 Hz string is made half as large, what would the new fundamental frequency be?

4. If you pluck a string 1/4 from one end, which partials will be missing? Which is likely to be the strongest one?

5. A viola string of length 60 cm is tuned to play a tone called G_3. Where should you place the finger on the fingerboard to play a tone one octave above G_3? One-fifth above G_3? One-fourth above G_3? (See Chapter 13 for the frequency ratios of these intervals.)

6. How does one reduce the stiffness of bass strings, which must be fairly thick? Acoustically, what is the effect of string stiffness? Is this effect desirable or not?

7. If you pluck a string 1/4 from one end and then touch it lightly at the midpoint, what modes will survive. Give the number of the modes.

▶ Chapter 18

Pipes

18.1 Pressure Pulse in a Pipe

You recall from the previous chapter that a standing wave is produced by the superposition of two identical waves that travel in opposite directions. In a string the opposing waves come from end reflections. Only at certain specific frequencies (or wavelengths) do these waves superimpose constructively to produce a mode or standing wave. At all other frequencies, no steady oscillation is possible because, having the wrong frequencies, after a few reflections the waves are out of step and interfere destructively.

The sound waves in a pipe are longitudinal pressure waves. An easy and quite correct way to visualize the sound wave in a pipe is to think of a Slinky (to keep it from sagging we suspend it from many fine threads). If you give the Slinky a push to compress it at one end (see Figure 14.5), the pulse travels along the Slinky much like a pressure pulse would travel along a pipe if the air is pushed at one end of the pipe. In Figure 18.1 we show the pressure pulse traveling to the right in two different ways: as a schematic picture of a pipe showing the air density by shading and as a graph showing the variation of air pressure along the pipe.

When the pressure pulse arrives at the end of the pipe, it is reflected because there is a sudden change on impedance (resistance) to air motion. At the open end, the compression pulse forces the air out of the pipe into the open, and inertia keeps the air moving out until a negative pressure in the pipe sucks the air back in, so in Figure 18.1 we show the reflected pulse as a *rarefaction* (a partial vacuum) instead of a compression.

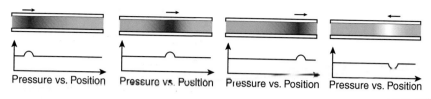

Pressure vs. Position   Pressure vs. Position   Pressure vs. Position   Pressure vs. Position

FIGURE 18.1 ▶ A pressure bump is traveling to the right in a pipe and is reflected at the open end.

Since the wave is reflected at the end of a pipe, under the right conditions standing waves can be formed just as in the case of a string.

18.2 Reflections in Open and Closed Pipes

The end of a pipe can be open, or it can be closed off with a stopper. Pipes closed at both ends are not useful as musical instruments because there is no way for the sound to come out or for the musician to set the air into oscillations—unless you think of a singer in a closed shower stall setting the air into oscillations!

Thus, usually by a *closed pipe* we mean a pipe **closed at *one* end,** while an *open pipe* is **open at *both* ends,** as shown in Figure 18.2. Unless we specify something different, it is understood that the inner hole (bore) of the pipe is cylindrical rather than, for example, cone shaped.

In the flute the mouthpiece represents an open end, even though it is off to the side, since air is free to escape. The clarinet has a cylindrical inner bore and can be considered a closed pipe, closed off by the reed.

It is reasonable to assume that the way the sound wave is reflected at the end of the pipe depends on whether the pipe is open or closed. At an open end, the air pressure cannot build up, since the air is free to escape, so there must be a pressure node at the open end. At a closed end, the stopper keeps the air from moving, so there must be a node of motion at the closed end. The physical conditions imposed on the air motion at the closed or the open ends are called "boundary conditions." For the example of the vibrating string, because the string is kept fixed at the ends, the boundary condition is that at both ends of the string there has to be a node of displacement.

Knowing the boundary conditions makes it easy to find out what modes of oscillation are possible, because the frequency or wavelength of the sound wave in the pipe must be just right to match the boundary conditions.

Cross section of an **open pipe**:

Cross section of a **closed pipe**:

FIGURE 18.2 ▶ Diagram of "open" and "closed" pipes.

18.2.1 Boundary Conditions

At the open end of pipe: pressure node, motion antinode (air escapes)
At closed end of pipe: motion node (air cannot move), pressure antinode (pressure belly)

It is a lot better to reason out this concept than to memorize it.

18.3 Standing Waves in Open Pipes

The wind player learns to set the air in the instrument into periodic oscillation by making periodic puffs of air in resonance with the modes of the pipe. It may seem implausible that a musician can make, for instance, exactly 440 puffs of air every second when she plays the standard A on the flute—exactly in step with the resonance frequency of the flute! How this is done will be discussed later. The mechanism involves a feedback between the flute oscillations and the oscillations of the musician's air stream.

Assume that the musician has set the air in a flute (open pipe) into steady oscillations. How do we picture the motion and pressure variations in the pipe? On the left of Figure 18.3, the dark and light areas show regions of compression and decompression of the air in the pipe during one oscillation of the fundamental mode. In the first picture the air is compressed most at the center. Air starts to flow out both ends of the pipe until the pressure is the same as the air pressure outside, but inertia keeps the air flowing out until the pressure inside the pipe is below normal air pressure. At this point the airflow reverses, and the air inside the pipe is compressed again.

On the right we show the corresponding graphs for air pressure and speed of airflow. It might seem surprising that the air is still ($v = 0$ throughout the pipe) at the moment when the pressure inside the pipe is the highest. But remember the pendulum motion: When the displacement is the largest (end of the swing), the speed is zero because at that moment the direction of travel reverses, while the speed is greatest at the middle of the swing when the displacement is zero. The same happens in the open pipe here: the air speed is greatest when the air is neither compressed nor diluted.

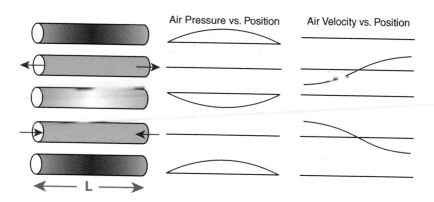

FIGURE 18.3 ▶ The fundamental mode of an open pipe. The dark and light areas of the pipe on the left indicate regions of high or low air pressure, respectively. The graphs on the right show the air pressure distribution along the pipe and the air velocity. Negative velocity means that the air moves to the left. Adjacent graphs are 1/4 of a period apart.

Note that in the fundamental mode, at the center of the pipe there is a node of motion. The air at the center gets compressed and decompressed, but it does not move! So the center is a node of air speed.

18.4 Fundamental Frequency of Open Pipe

Just as in the case of the string, the frequency of excitation (external pushes) must correspond to the round-trip travel time of the wave. For a pipe, the speed of travel is the speed of sound in air, $v = 344$ m/s at room temperature. Thus, the fundamental frequency of an open pipe of length L is

> Fundamental frequency of open pipe: $f_1 = \dfrac{v}{2L}$

Here, the appropriate value for v to use is the speed of sound in the gas that fills the pipe (room air), because it is the oscillation of the air rather than of the pipe wall material that produces the standing wave. The same applies to all wind instruments. Note that the pipe diameter does not appear in the formula; the diameter of the pipe has almost no effect on the frequency of the pipe's tones. For a string, the diameter matters but not for a pipe. You might wonder why that is, because, after all, for a pipe the mass of air that is oscillating is greater too when the pipe has a larger diameter. But the difference is that for a pipe of larger diameter, the push on the air by the propagating sound wave is also bigger because the air pressure acts over a larger area. The larger push offsets the larger mass. For a string, too, if you increased the diameter *and* the tension, the frequency would stay the same.

A string of length L can produce any frequency because we can change the mass of the string or the tension at will. But for a pipe of length L, the frequency is fixed because we cannot change the speed of sound in air! So when a string player tunes the instrument, she changes the tension, but when a flutist tunes her instrument, she makes it a little shorter or longer by pulling out the mouthpiece from the body of the flute.

Example: A flute (with all finger holes closed) plays the tone C, which has a frequency of 260 Hz. How long is the flute?

Answer: Solve $f_1 = v/2L$ for L:

$$L = \frac{v}{2f_1} = \frac{344 \frac{m}{sec}}{2 \times 260 \frac{1}{sec}} = 0.66m$$

18.5   Higher Modes of Open Pipe

Because modes are sinusoidal oscillations, to find the higher modes, we must find those *sinusoidal standing waves* inside the open pipe with pressure nodes at each end of the pipe (boundary condition).

For the **second mode** there will now be **two half-wavelengths inside the pipe**, again with a pressure node at both ends. From Figure 18.4 we see that for the second mode there is a region of high pressure and a region of low

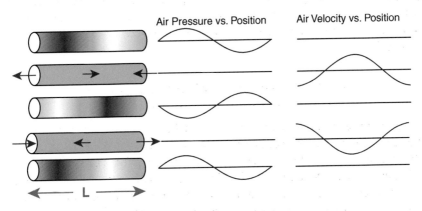

FIGURE 18.4 ▶ Second mode of an open pipe. The dark and light areas of the pipe on the left indicate regions of high and low air pressure. The graphs on the right show the air pressure distribution along the pipe and the air speed. Note that when the air flows out of the left end it flows into the right end, which is opposite to the fundamental mode of Figure 18.3. Adjacent graphs are 1/4 of a period apart.

pressure inside the pipe. Initially, air flows out of the left side and into the right side of the pipe, which is opposite to what we found for the first mode. Note that in this mode the air at the center of the pipe moves back and forth, while for the fundamental mode the air in the center never moves.

For the next higher modes, three, four, five, etc half-wavelengths fit inside the length of the pipe. We summarize the pressure distributions for all modes in Figure 18.5, where we mark the pressure nodes (places of constant pressure) with a green arrow and the antinodes with a red arrow. Remember that the air velocity is opposite: *The air velocity is largest where the pressure has a node and smallest where the pressure has an antinode.*

As you can see from Figure 18.5, the higher the mode, the shorter the wavelength, and therefore the higher the frequency. But all frequencies are multiples of the first mode frequency: They are **harmonic**. This is very important from the musical standpoint. The fact that the modes of pipes and strings are harmonic is responsible for the pleasing sound of the instruments.

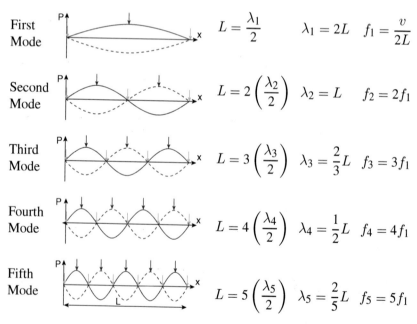

First Mode $\qquad L = \dfrac{\lambda_1}{2} \qquad \lambda_1 = 2L \quad f_1 = \dfrac{v}{2L}$

Second Mode $\qquad L = 2\left(\dfrac{\lambda_2}{2}\right) \quad \lambda_2 = L \qquad f_2 = 2f_1$

Third Mode $\qquad L = 3\left(\dfrac{\lambda_3}{2}\right) \quad \lambda_3 = \dfrac{2}{3}L \quad f_3 = 3f_1$

Fourth Mode $\qquad L = 4\left(\dfrac{\lambda_4}{2}\right) \quad \lambda_4 = \dfrac{1}{2}L \quad f_4 = 4f_1$

Fifth Mode $\qquad L = 5\left(\dfrac{\lambda_5}{2}\right) \quad \lambda_5 = \dfrac{2}{5}L \quad f_5 = 5f_1$

FIGURE 18.5 ▶ Each graph is a graph of pressure versus position inside the pipe. Red arrows show pressure antinodes, and green arrows show pressure nodes. Notice that in odd harmonics there is a pressure antinode at the center of the pipe, while in even harmonics there is a node at the center of the pipe.

18.6   Fundamental Frequency of Closed Pipe

As mentioned in section 18.2, for a *closed pipe* (closed at one end), there must be a node of pressure at the open end but an *antinode of pressure at the closed end*. At the closed end the air pushes against the stopper so that the air pressure varies at the closed end, while at the open end the air is free to escape and the pressure is constant. The pressure and air velocity graphs for a closed pipe are shown in Figure 18.6.

You may have noticed in Figure 18.6 that for the fundamental oscillations of the closed pipe, 1/4 of a wavelength fits inside the pipe of length L, while for the open pipe the wavelength was half as long, so 1/2 a wavelength fits. In other words, considering two pipes of the same length, the closed pipe has twice the wavelength of the open pipe. Longer wavelength means lower frequency: *A closed pipe has half the frequency of an open pipe of the same length.*

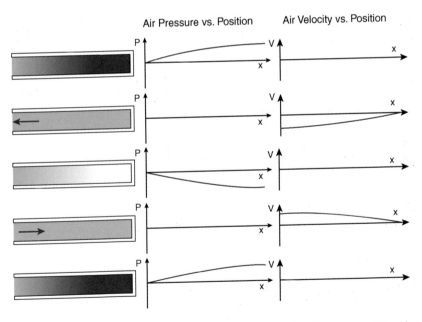

FIGURE 18.6 ▶ Fundamental mode of closed pipe. The dark and light areas of the pipe on the left indicate regions of high or low air pressure. The graphs on the right show the air pressure distribution along the pipe and the air velocity. Negative velocity means that the air moves to the left. Adjacent graphs are 1/4 of a period apart.

Here it is expressed as an equation:

$$\boxed{\textbf{Fundamental of closed pipe: } f_1 = \frac{v}{4L}}$$

A particularly attentive reader may argue that if we again were to use the argument that the round-trip time of a pulse should equal the period of the fundamental, we would get the same answer for the period no matter whether the pipe is open or closed. This apparently is not the case. Why not? The answer is that at the closed end a positive pulse is reflected as a positive one. Thus, a belly of the sine wave is reflected at the closed end as a belly, meaning that the phase of the wave jumps by 180°. So we lose a half-wavelength in the reflection at the closed end.

Example: A 2-m-long organ pipe is closed at one end. Find the fundamental frequency.

Answer:

$$f_1 = \frac{v}{4L} = \frac{334 \text{ m/s}}{4 \times 2 \text{ m}} = 43 \text{ Hz}.$$

On a pipe organ, the pipes of lowest frequency are often closed pipes. To get the same 43 Hz fundamental from an open pipe requires a pipe twice as long—that is, 4 m, or about 12 feet.

18.7 Higher Modes of Closed Pipe

For the next higher mode, we must again be sure that there is a pressure node at the open end and a pressure antinode (belly) at the closed end. The picture for the second mode is shown in Figure 18.7.

So what is the frequency f_2 of the second mode? There are now **three** (1/4) waves inside instead of just one, so the wavelength is three times shorter and the frequency **three times higher** than the fundamental: $f_2 = 3f_1$. In contrast to strings and open pipes, the next mode does **not** have twice the frequency!

Following we show the pressure distribution for the first five modes of the closed pipe. To always have an antinode (belly) at the closed end, we must use an **odd multiple of** $\lambda/4$ inside the pipe—namely, 1, 3, 5, 7, and so on. Correspondingly, the frequencies of the modes are 1, 3, 5, 7, 9 times the fundamental frequency, not 1, 2, 3 times the fundamental!

$$\boxed{\textbf{Closed pipe: } f_n = (2n-1) \times \frac{v}{4L}, \text{ where } n = 1, 2, 3, \dots}$$

Figure 18.8 shows the pressure distributions for the first five modes of the closed pipe.

Air Pressure vs. Position   Air Velocity vs. Position

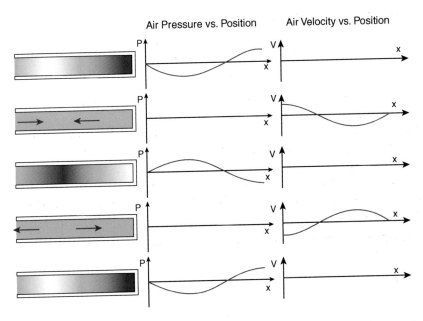

FIGURE 18.7 ▶ A closed pipe oscillates in the second mode. The pressure has a node at the open end and an antinode at the closed end. The air velocity has an antinode at the open end (air rushes in and out) but a node at the closed end because the stopper keeps the air at the closed end from moving. The length of the pipe is $L = (3/4)\lambda$.

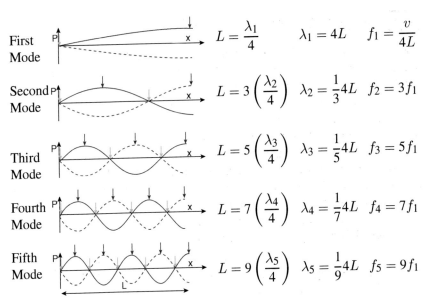

First Mode $\qquad L = \dfrac{\lambda_1}{4} \qquad \lambda_1 = 4L \qquad f_1 = \dfrac{v}{4L}$

Second Mode $\qquad L = 3\left(\dfrac{\lambda_2}{4}\right) \qquad \lambda_2 = \dfrac{1}{3}4L \qquad f_2 = 3f_1$

Third Mode $\qquad L = 5\left(\dfrac{\lambda_3}{4}\right) \qquad \lambda_3 = \dfrac{1}{5}4L \qquad f_3 = 5f_1$

Fourth Mode $\qquad L = 7\left(\dfrac{\lambda_4}{4}\right) \qquad \lambda_4 = \dfrac{1}{7}4L \qquad f_4 = 7f_1$

Fifth Mode $\qquad L = 9\left(\dfrac{\lambda_5}{4}\right) \qquad \lambda_5 = \dfrac{1}{9}4L \qquad f_5 = 9f_1$

FIGURE 18.8 ▶ First five modes of closed pipe. The graphs show pressure versus position inside the pipe. Red arrows show antinodes of pressure, and green arrows show nodes of pressure.

Example: Let us find the first three mode frequencies of a 2-m-long closed pipe.

Answer: We found that the fundamental of a 2-m-long closed pipe is 43 Hz. The next two modes have frequencies of 3×43 Hz $= 129$ Hz and 5×43 Hz $= 215$ Hz.

18.8 Playing Tunes on Wind Instruments: Fingerholes and Overblowing

Just as for the string, in wind instruments the player does not excite only a single mode but many modes at the same time. The sound from an open pipe is the sum (superposition) of many sinusoidal oscillations of frequency f_1 and multiples of f_1, whereas for the closed pipe only the odd multiples of f_1 are present.

String players can easily change the fundamental frequency simply by putting a finger on the fingerboard to make the string shorter. Wind players have to make their pipe shorter or longer by opening different finger holes where the air can escape and make a pressure node somewhere along the pipe or by making the pipe longer or shorter with valves (trumpet) or a slide (trombone).

But it would be difficult to cover a very large range in frequency this way. Every octave requires making the pipe half as long so after three octaves it is 1/8 the original length and there is almost nothing left—not to mention the tremendous number of finger holes the musician would have to open and close. Another way to produce frequency changes in wind instruments is to "overblow" to the next mode: the wind player learns to blow in such a way that the excitation frequency of an open pipe jumps to the second mode, so the full length of the pipe plays $2f_1$—that is, the octave. Or it could be even three or four times the fundamental. One striking application of this is the bugle, which is a trumpet without any valves. The bugle can nevertheless play a sequence of tones: $f, 2f, 3f, 4f$, and so on, called the harmonic series or the harmonic scale. By what magic the musician can at will excite different modes, we will discuss in Chapter 21.

18.9 Other Shapes

Except for the flute and some organ pipes, wind instruments are not usually simple cylinders. Brass instruments have flares attached to the end, not only to improve sound radiation but also to shift the resonance frequencies. Some instruments (oboe, bassoon) have a conical rather than a cylindrical opening (see Chapter 21).

The math for figuring out the modes of these instruments is complicated, but we want to mention one interesting fact. The modes of a conical pipe are for some reason mathematically identical to those for an *open* cylindrical pipe of the same length. Of course, for a conical pipe we imagine a mouthpiece attached to the tip of the cone for the musician to blow so that the air inside is set into oscillation.

18.10   Acoustic Length

In connection with the open pipe equation, we mentioned that the diameter of the pipe does not appear in the frequency formula—in other words, the frequency is the same no matter how large a diameter the pipe has. But this is an oversimplification: In actual fact, the pipe oscillations extend beyond the physical end of the pipe by some fraction of the pipe diameter. The effect of this extra length is to slightly lower the frequencies of all the modes. The "acoustic length" of a pipe is the value of L in the equation that corresponds to the **actual** frequency of the pipe. The amount by which the acoustic length exceeds the physical length is about 0.6 times the radius of the pipe.

18.11   Questions

Assume speed of sound in air 344 m/s.

1. How does the frequency of an 80-cm-long open pipe change if you:
   (a) change the length from 80 cm to 100 cm? (Find the frequency for both lengths.)
   (b) keep the length at 80 cm but close one end of the pipe?
   (c) change the diameter of the pipe from 2 cm to 4 cm?
   (d) increase the air temperature?

2. (a) The flute is a cylindrical pipe of length 66 cm, open at both ends. Find the fundamental frequency of the flute when all finger holes (tone holes) are closed.
   (b) The clarinet has almost the same length as the flute, but the fundamental frequency is about half that of the flute. Is the clarinet an open or a closed pipe?

3. The fundamental frequency of an open organ pipe is 200 Hz.
   (a) What is the length of the pipe?
   (b) What is the frequency of the same pipe if one end is closed?

4. Wind instruments usually have finger holes or valves so that one can change the effective length of the instrument. Besides changing the length, what other way is there to change the pitch of a wind instrument?

5. How long an organ pipe would you need to play a 40 Hz tone?

6. (a) Find the fundamental frequency and the frequencies of the first two overtones of an open pipe of 60 cm length.

   (b) If the same pipe is closed at one end, what are the corresponding frequencies?

7. (a) Make a graph of the pressure at different instances in a cylindrical pipe open at both ends when it oscillates in the fundamental mode, and

   (b) in a cylindrical pipe closed at the left end.

8. Between room temperature (20°C) and body temperature (37°C) the speed of sound increases by 10 m/s. A flute has a frequency of 260 Hz when it is cold. Find the frequency when the flute is warmed to body temperature by the flutist's breath. (Hint: Use proportions to relate frequencies to speed of sound. What is the ratio of speed of sound at the two temperatures? What is the frequency ratio?)

Chapter 19

Fourier Analysis

The Fourier Theorem

When we looked at modes of a string, we represented the intensity of the various modes by a bar graph, which showed the amplitude of all the modes that make up the complex vibrations of the string. A bar graph, of course, is much easier to draw than the complex wave, which is the superposition of the various sine curves. You may remember from the examples in Chapter 13 how tedious it is to draw the actual wave shape when two oscillations are superimposed, not to mention three or four!

Looking at a wave shape—for instance, of the human voice—on an oscilloscope may be fun but it is very hard to interpret or to identify which feature of the wave shape is different when you compare the voice of an opera star to your own voice. A young singer might decide to improve her voice by trying to make her oscilloscope trace look similar to the oscilloscope trace from the recording of a famous star, but this would be futile because the oscillations are so complex. In addition, changing the phase (the timing) between oscillations changes the wave shape but not the tone we hear, which is another indication that the wave shape is not all that important. Rather, it is the intensities of the partials—and not their relative timing—that determine the tone quality.

Instead of dealing with the complicated wave shapes that voices and instruments produce, *Fourier analysis* allows us to describe a steady tone of any complexity by just a few numbers or by a simple graph.

Physics in the Arts, Revised Edition

Baron Jean Baptiste Joseph Fourier, a French mathematician (1768–1830), presented a mathematical proof of the *Fourier theorem*. This theorem has three parts:

1. Any periodic oscillation of period T can be represented as a sum of sine waves with appropriate frequencies, phases, and amplitudes.
2. Only certain frequencies are necessary—namely, those that are multiples of the fundamental frequency $f = 1/T$.
3. There is only one combination of sinusoidal oscillations that produces the given complex wave.

In other words, the Fourier theorem states the following:

Any periodic oscillation of period T can be expressed as a sum of sine curves whose frequencies are multiples of f = 1/T and whose amplitude and phase are suitably chosen.

For example, if a singer's voice makes a very complex oscillation of some repetition frequency f, this oscillation can be represented by the sum of *Fourier components*—namely, sinusoidal oscillations of frequencies f, $2f$, $3f$, and so on. Of course, for a string or a pipe we are not surprised that this is possible because we have learned from the physics of their oscillations that the tone they produce is a superposition of modes of frequency $f, 2f, 3f$, and so on. Musicians call the Fourier components the *partials* of the tone.

Why are the frequencies of the Fourier components exact multiples of the fundamental frequency? If the fundamental oscillation has a period T, the periods of the higher oscillations must fit into T once, or twice, or three times, and so on. If they did not, they would spoil the periodicity. Thus, we readily understand that the frequencies of the Fourier components must be multiples of the fundamental frequency. It may be helpful to know that often the Fourier theorem is formulated differently. Instead of specifying amplitude and phase for each Fourier component, one often specifies each Fourier component as the sum of a sine and a cosine oscillation of the same frequency, $(a \cdot \sin + b \cdot \cos)$, where a and b are adjusted to fit the complex oscillation. Readers who know trigonometry may have learned that depending on the relative magnitude of a and b, the resulting oscillation shifts in phase.

19.2 Sound Spectrum

The Fourier spectrum of a musical tone (also called the **sound spectrum**) is a graph that represents the amplitude of the various partials of a complex tone versus the frequency of the partial. Often, if there are lots of partials, we may show the number of each partial instead of the actual frequency, but we still need to specify the frequency of the fundamental.

How does one find the Fourier components (partials) of a complex tone? There are two ways: You can record the exact wave shape and use a mathematical procedure involving integral calculus to find the amplitudes of the partials. But for studies in acoustics, a much more convenient and elegant approach is to use a "Fourier analyzer," which carries out this calculation and immediately displays the spectrum of sound picked up by a microphone on a computer screen. Any present-day desktop computer with a sound card and either an internal or external microphone can be used as a Fourier analyzer. The necessary programs are now readily available for free or for a small fee on the Internet. If you do a search using the keywords "Fourier analyzer," you can have great fun exploring your voice or a musical instrument.

To get used to the concepts, the figures in this chapter show some periodic oscillations and the corresponding Fourier spectrum. We first deal with calculated wave shapes (Figures 19.1 to 19.4) and later show examples of actual measurements.

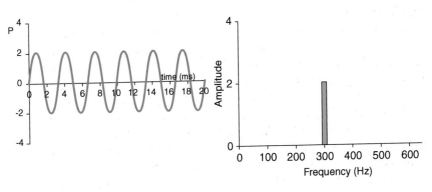

FIGURE 19.1 ▶ Pressure variations and corresponding sound spectrum of a pure tone of frequency 300 Hz and amplitude 2 units.

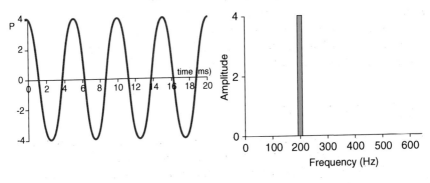

FIGURE 19.2 ▶ Pressure variations and corresponding sound spectrum of a pure tone of frequency 200 Hz and amplitude 4 units.

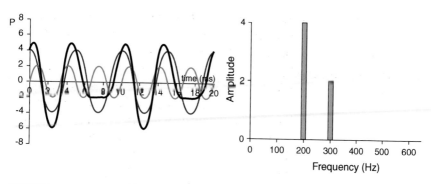

FIGURE 19.3 ▶ Pressure variations and corresponding sound spectrum when the two tones of Figures 19.1 and 19.2 are played at the same time.

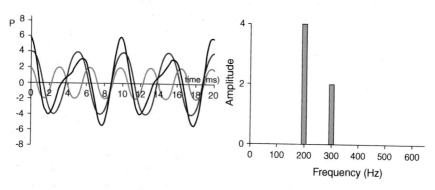

FIGURE 19.4 ▶ Pressure variations and corresponding sound spectrum for the same two tones of Figure 19.3 but shifted in relative timing (phase). The pressure variations are different, but the sound spectrum and the perceived sound stays the same.

1. **A pure tone (e.g., tuning fork):** The left side of Figure 19.1 shows the sound pressure variations. The right shows the corresponding Fourier spectrum. It is obviously a lot easier and more instructive to draw the Fourier spectrum than the actual air pressure variations. The figure tells us that there is one sine wave present of frequency 300 Hz and amplitude 2 units.

2. A pure tone of different phase, frequency and amplitude (Figure 19.2).

3. **A complex tone consisting of two partials:** In Figure 19.3 we have added the preceding two oscillations to produce a more complex tone. The period of the resulting oscillation is 10 ms, which corresponds to a 100 Hz frequency.

    Now it is even more obvious that the Fourier spectrum is an elegant way to see what is happening. If you only were given the complex oscillation (black curve in Figure 19.3), you would have a hard time figuring out its components, while the Fourier spectrum immediately tells you what partials made up the black curve.

The missing fundamental: In the preceding example, the complex oscillation clearly has a frequency of 100 Hz—in other words, it repeats itself every 10 ms, or 100 times per second. As we explained earlier, this frequency is the largest common divisor of 200 Hz and 300 Hz, but what is the amplitude of the fundamental in the Fourier spectrum? The fundamental has amplitude 0 (zero)—that is, there is no fundamental present. Yet, 100 Hz is the fundamental in the sense that the repetition frequency is 100 Hz. We refer to this as the *missing fundamental*.

It is interesting to ask ourselves what frequency we actually hear in this case. If you have several partials, but the fundamental is not actually present in the complex oscillation, the pitch still sounds like the fundamental that is not there! This is one reason why a small radio on your nightstand that has virtually no bass response still can reproduce bass sounds to some extent, because the higher partials of the actual bass suggest to the ear-brain system the presence of the fundamental, even if the speaker in the radio does not reproduce it.

4. **Change in phase affects wave shape but not Fourier spectrum:** Drawing the spectrum instead of the wave shape is a lot easier and more instructive, but here is a possible problem: when you change the timing (phase) between the oscillations, the wave shape changes considerably, as shown in Figure 19.4. But the spectrum graph stays the same because the graph only says how much of each oscillation there is, not how they are timed. So how can one say that the spectrum graph represents the real sound?

It is indeed sufficient to show only the amplitude but not the phase of the partials because we cannot hear differences in phase between the partials! Two oscillations (e.g., left-hand side of Figures 19.3 and 19.4) can have rather different wave forms yet sound exactly the same. We say that *the ear is insensitive to the phase of the partials*. Stated another way, the human hearing apparatus is a Fourier analyzer that takes apart the complex wave shape into its components (partials) without paying attention to the detailed wave shape. This is rather surprising in view of the coding of sound in nerve pulses, the distribution of which in fact is sensitive to phase, so it is only later, in the brain, that the phase information is ignored.

This raises another question: when we listen to a complex tone from a musical instrument, do we hear that the tone is made up of partials? The answer is no. We integrate all the partials into one perception of one tone, with one frequency. In our perception, all the partials fuse into a single tone, which we identify as a flute, a violin etc. This process—fusion of partials—happens in the brain and is poorly understood. Some musicians claim that they can hear partials but most of us don't.

5. **The sound spectrum of musical instruments.** In Figure 19.5 we show the sound spectra for two instruments, an alto saxophone and a clarinet. This illustrates that real instruments usually have many partials. A violin would have even more. Second, it illustrates that the clarinet has strong odd and weak even partials (at least for the first few partials), similar to

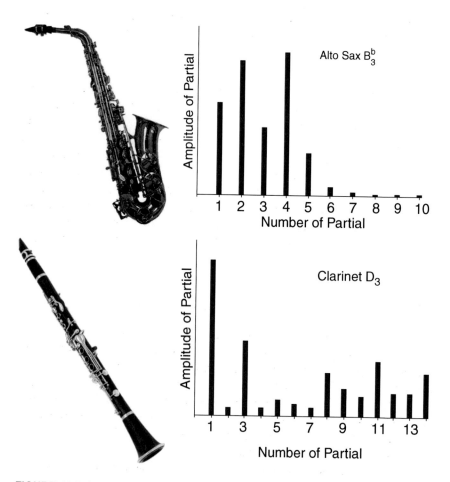

FIGURE 19.5 ▶ Sound spectrum of an alto saxophone playing a tone with frequency 233 Hz and a clarinet playing a frequency of 134 Hz. The clarinet has strong **odd partials**, while the saxophone has odd and even partials of similar strength.

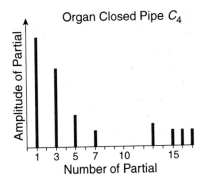
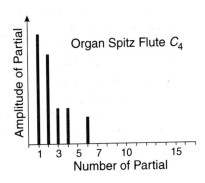

FIGURE 19.6 ▶ Sound spectrum of a closed organ pipe (left) and a conical organ pipe (right), both playing a frequency of 262 Hz. Note that the closed pipe has only odd partials.

a closed pipe, while the saxophone has strong even and odd partials. We will discuss that more when we talk about instruments in Chapter 21.

Figure 19.6 shows the spectrum of two organ pipes playing the same tone C_4, which has a frequency of 262 Hz. The names of organ pipes are often derived from German names: the name *gedackt* pipe comes from *gedeckt*, meaning "covered"—that is, a closed pipe (left graph). It shows odd harmonics only. In contrast, the *Spitzflöte* (*spitz*, meaning "point"— that is, a pointed, conical pipe) has odd and even harmonics. The two pipes have the **same pitch but very different timbre:** The Spitzflöte sounds very nasal, while the closed pipe has a smoother, rounded tone.

19.3 Fourier Analyzer (Sound Analyzer)

Almost any computer with a sound card and a microphone can be used to record and to graph almost instantaneously the sound spectrum of a tone played or sung into the microphone. Such displays are now used in voice training. Experienced singers produce a vowel sound in such a way that all partials are immediately present at the beginning of the tone, while for less experienced singers the higher partials come into play only slowly. By watching the spectrum on the computer screen, the singer learns how to avoid this undesirable "blooming" of the partials.

Two examples of sound spectra acquired with a computer are shown in Figures 19.7 and 19.8.

In Figure 19.7 one of the authors was singing "aah" as in "father" at about 220 Hz. He actually tried to match the frequency of a 440 Hz tuning fork but

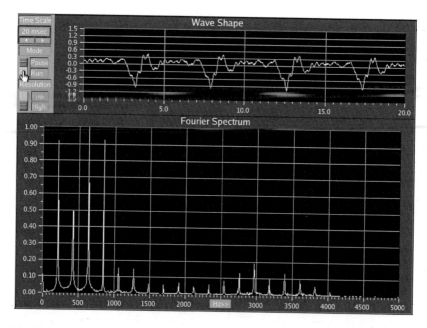

FIGURE 19.7 ► Fourier analysis done on a computer. The example shows Willy's voice when he sings a sustained "aah" sound at 220 Hz. The computer displays the sound pressure variations detected by a microphone (top part of the computer screen) and the sound spectrum (bottom part of the screen).

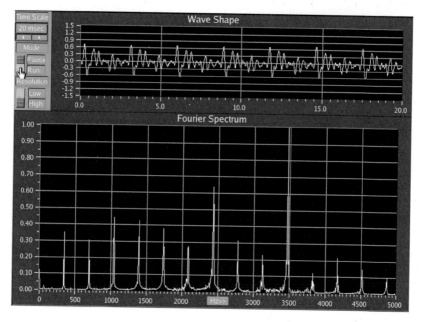

FIGURE 19.8 ► Sound spectrum of a violin playing a frequency of 350 Hz. Note that the tenth partial is much stronger than the fundamental.

sang the tone an octave too low. It is interesting that even good singers, when they try to match the tone of an instrument, often find that they sing in a different octave.

The top panel of the computer screen reproduced in Figure 19.7 shows the air pressure variations picked up by a microphone. Note that the wave is periodic but quite complicated. The time scale shows that the period is approximately 5 ms. The panel below shows that the sound spectrum contains partials up to the eighteenth partial. There are stronger groups in the region up to about 1000 Hz, and then again around 3000 Hz. These regions are called the **formants** of the vowel "aah" in Willy's voice and are the result of broad resonances in the throat and oral cavity. Notice the difference between Willy's voice and a violin by comparing Figures 19.7 and 19.8.

19.4 Fourier Synthesis

If any periodic oscillation can be represented as a sum of sine waves, it must be possible to artificially synthesize any periodic oscillation by adding sine oscillations. Accordingly, a violin tone or Willy singing "aah" can be synthesized by adding many sine waves. Electronic synthesizers produce musical tones resembling real instruments in this way, but a more interesting use of such synthesizers is to produce new and different sounds.

As an illustration of Fourier synthesis, let's see if we can synthesize a square wave (black curve in Figure 19.9) with a combination of sine curves. We start by making a fundamental sinusoidal oscillation of frequency f that resembles the square wave as much as possible (red sine wave in Figure 19.9). The next higher oscillation will have twice the frequency of the fundamental (cyan sine). But you can see that adding such an oscillation is a bad idea because the square wave is mirror-symmetric left and right of the vertical dashed lines, while the oscillation with frequency $2f$ is not symmetric. Thus, we should make the amplitude of the $2f$ oscillation zero. Adding $3f$, on the other hand, really helps (Figure 19.10). Adding a bit of $3f$ to the fundamental already makes the sum of these two a lot closer to the square (purple line in Figure 19.10).

Figure 19.11 shows that the sum of four harmonics ($f, 3f, 5f, 7f$) is getting closer to the square wave, but note that the steep vertical sides of the square wave cannot really be made up of low-frequency sine waves. To produce a vertical step would require infinitely high partials. Finally, Figure 19.12 shows that 51 partials are a pretty good representation of the square wave.

It is generally true that representing a very rapid rise or fall of the curve requires very high frequencies.

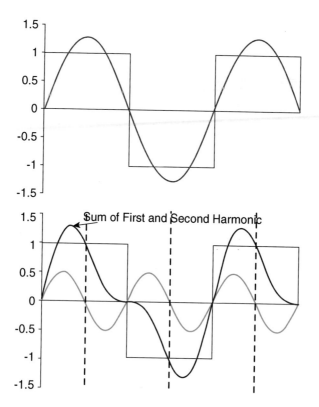

FIGURE 19.9 ▶ Top: A square wave (black curve) is approximated by the red sine wave. Bottom: The square wave is mirror-symmetric about the dashed vertical lines. Adding an oscillation of twice the frequency would destroy the symmetry.

19.5 Why Can't We Synthesize a Stradivari?

Can we synthesize the tones musicians play? Indeed, it is easy enough to synthesize a perfectly steady, perfectly periodic oscillation, but this is not what a musician aspires to play. In the first place, a musical tone is not a start-play-stop oscillation. Even though we are not consciously aware of it, the tone builds up over a fraction of a second, and at the end, it does not stop abruptly but takes a certain time to decay. The characteristic attack and decay of a tone are both important. As mentioned in Section 12.3, if a steady tone of an instrument is recorded and then replayed with the beginning and the end of the tone cut off abruptly, the instrument looses so much of its character that in some cases it becomes nearly unrecognizable. For a synthesizer to reproduce

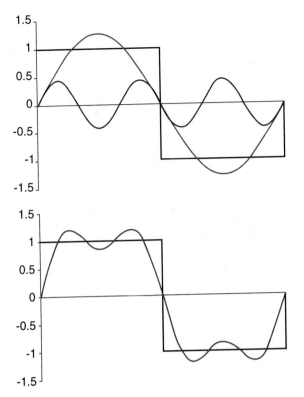

FIGURE 19.10 ▶ Top: Here we added an oscillation of frequency $3f$ to the fundamental of frequency f. Bottom: The sum of these two shows that we are already getting closer to representing a square wave (lower graph).

the attack and decay, one must keep in mind that each of the approximately 20 partials has a different rise and decay time, which adds a lot of complexity.

But a more fundamental problem is that real musical tones are not completely periodic. Take a violinist as an example. As she plays a tone, she has several parameters she can vary with time during the tone: speed of bow, pressure on bow, distance of bow from bridge, finger position (vibrato). But each time she plays the same tone again, she will choose slightly different parameters, and it is these choices that distinguish artists from one another even if they were to play the same Stradivari! So no, there is no point trying to synthesize the beautiful, lively, varied tones musicians play be it on a Stradivari or any other instrument. The main purpose of synthesizers then is to produce interesting new sounds, different from actual instruments.

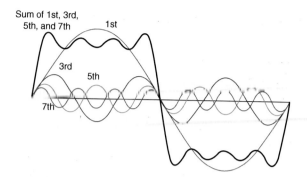

Sum of 1st, 3rd,
5th, and 7th   1st

3rd

5th

7th

FIGURE 19.11 ▼
Adding more waves of
the proper frequency
(odd multiples only)
and the right ampli-
tude makes the result
closer and closer to a
square wave.

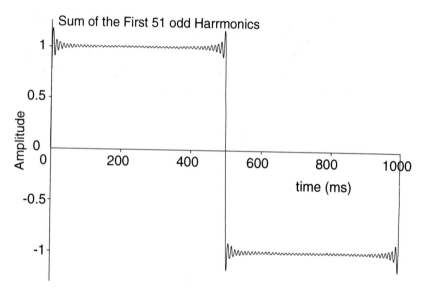

FIGURE 19.12 ▶   The sum of 51 odd harmonics shows a very good approximation of a
square wave.

19.6   Questions

NOTE: We usually cannot figure out the amplitudes of the overtones but can
only deduce which overtones are present and what their frequencies are. Thus,
when you are asked to draw a Fourier spectrum, the Fourier components
should be in the right position, but their intensity is arbitrary.

1. Suppose you have a complicated but periodic oscillation of period 5 ms.
    What statement did Fourier make about representing this complicated
    wave as a superposition of sine waves?

Answer: Any complex oscillation of period T (e.g., 5 ms) can be represented

as _____

whose frequencies are _____

For this example, the frequencies would be _____, _____,

and _____.

2. Draw the Fourier spectrum of a string whose fundamental frequency is 200 Hz when it is plucked at the midpoint.

3. (a) Draw the Fourier spectrum of a closed pipe with fundamental frequency 300 Hz. (b) Draw the spectrum when the same pipe is open at both ends.

4. A complex tone from a violin has a frequency of 300 Hz. Which partials are present in the complex wave shape?

5. What might the Fourier spectrum of a 500 Hz violin string look like when it is plucked 1/3 the length from one end? (String players refer to plucking the string as "playing pizzicato")

Chapter 20

Musical
Scales

A musical scale is a sequence of frequencies, most often arranged in ascending order. A scale normally is contained within one octave—that is, within a frequency range of 2:1. The tones of the scale are also referred to as the *notes* of the scale. The term *note* is used either to denote the symbols that represent the tones of the scale on paper or the tones themselves.

Why do we need a scale? To make music in an organized fashion, one needs some set of frequencies that people (more or less) adhere to, at least within a given culture. Otherwise, how would one tune a piano and bore holes in the body of wind instruments and so forth? And how would an entire chorus sing "in tune"?

We could simply describe the Western scale by a table of frequencies and could say that's the way the scale developed over the centuries. But physics tries to understand why it is that way. *Why* does the piano keyboard end up with the funny arrangement of black and white keys where a black key is missing between some of the white keys? It is clear that a scale evolves over centuries, based on the aesthetic judgment of generations of musicians, but here we seek to find the reason in particular for the development of the scale used in most Western music. We want to show that quite likely the scale as we know it is a direct result of the desire to make a scale that allows for harmonious frequency ratios. Some of it will involve some tedious calculations, but the reader who is only interested in the end result can skip over it.

Physics in the Arts, Revised Edition
© 2012 Elsevier Inc. All rights reserved.

20.1 | Musical Intervals

When you sing a tune, your voice produces a sequence of musical tones of different pitch—that is, a sequence of sounds with different frequencies. As a child, if you grew up in the United States or in Europe, you may have learned the song "Are You Sleeping, Are You Sleeping, Brother John?" (or "Frère Jacques, Frère Jacques, Dormez-Vous?," or "Fra Martino, Campanaro, Dormi Tu?," or "Bruder Jakob, Bruder Jakob, Schläfst Du Noch"). Frequency measurements would show the following:

Are	you	sleep-	-ing
150 Hz	168.5 Hz	187.5 Hz	150 Hz

But this is not the only way one can sing this tune: A woman may sing it at a higher pitch:

Are	you	sleep-	-ing
200 Hz	225 Hz	250 Hz	200 Hz

or a man with a low voice may sing it like this:

Are	you	sleep-	-ing
100 Hz	112.5 Hz	125 Hz	100 Hz

These different sequences of tones to us are "the same tune"—that is, the same sequence of musical intervals. The question is, what about these frequencies is the *same* for all three of the preceding sequences of numbers?

The answer is simple. ***The ratios of the various frequencies are the same.*** The frequency of the second tone divided by that of the first in each case is the same—namely, 9/8. Let us check:

$$\frac{112.5}{100} = 1.125 = \frac{9}{8}; \quad \frac{168.75}{150} = 1.125 = \frac{9}{8}; \quad \frac{225}{200} = 1.125 = \frac{9}{8}$$

The ratio between the next two tones, "you" and "sleep-," is also the same for all three renditions of the tune.

We must therefore conclude that what sounds to us like the same musical interval, in physical terms means the same frequency ratio. Note that the frequency differences in Hz are quite different; it is the *ratio* that matters.

A musical interval is characterized by the ratio of the two frequencies. Intervals sound the same if they have the same frequency ratios.

This means that if you divide (or multiply) all frequencies of a tune by the same number, it will again be the same tune. On some old record players you can change the speed of the turntable. If you change the speed, the music, of course, is faster or slower, but it is exactly the same tune in a higher or lower pitch. All the frequencies are multiplied by the same number.

Examples of musical intervals (the origin of the names is explained later):

A **2:1** ratio is called the *octave*: The interval from 200 Hz to 400 Hz sounds the same as the interval from 300 Hz to 600 Hz.

A **3:2** ratio is called a *fifth*: The interval from 200 Hz to 300 Hz sounds the same as the interval from 300 Hz to 450 Hz.

A **5:4** ratio is called a *major third*: In the preceding example ("Are you sleeping ..."), the third tone ("sleep-") is a major third above the first ("Are").

A **6:5** ratio is called a *minor third*.

Note that a 3:2 ratio (3/2 = 1.5) is the same interval as the inverse 2:3 ratio (2/3 = 0.666). The only difference is that in one case we "go up one fifth," in the other case we "go down one fifth."

Example: What is the frequency of a tone one fifth above A = 440 Hz? A tone one fifth below?

Answer: Going up means to multiply by a 3/2. Thus, the frequency becomes $440 \times (3/2) = 660$ Hz.

Going down means either to **divide by 3/2 or multiply by 2/3**. Multiplication by some fraction like 2/3 is the same as division by the reciprocal 3/2:

$$440 \times \frac{2}{3} = \frac{440}{3/2} = 293.3 \text{ Hz}$$

20.2   Consonance (Harmony): Simple Number Ratios

In all musically developed cultures, tunes are played with a number of tones that have well-defined intervals. These are the intervals of the musical scale. The tuning of the piano or the placement of frets on the guitar are chosen to correspond to those particular intervals. But what exactly are those intervals, and how were they chosen to start with?

The establishment of a musical scale is a cultural development that extends over centuries and involves subjective preferences. However, one finds that all cultures seem to have a preference for certain musical intervals (frequency ratios), like the octave, the fifth and other simple-number ratios. The preference for simple-number ratios quite likely is related to the fact that in most music two or more tones are heard simultaneously. Compositions in which two or more tones are sounded simultaneously are called "polyphonic" (from the Greek *poli,* meaning "many" and *phonos,* meaning "sound"). Not all tones played in combination simultaneously sound pleasing to the ear. Some intervals are "rough" or "dissonant," while other intervals are particularly pleasing, harmonious or "consonant." But what combinations of frequencies $f_1, f_2, \ldots$ are in harmony, and why? A plausible explanation is the following:

Simple-number frequency ratios are consonant ("in harmony") because the various partials of the two tones being played simultaneously either coincide exactly or are different enough to avoid beats or roughness.

Roughness comes about when two frequencies are near one another but not so close that one discerns the beats individually (e.g., 400 Hz and 410 Hz). For example, two tones a fifth apart, such as 200 Hz and 300 Hz are consonant because the partials of one (200, 400, 600, 800, 1000, 1200 Hz) and of the other (300, 600, 900, 1200 Hz) have a particular relationship: Some partials coincide exactly (here 600, 1200 Hz), while the others are so different that they produce no roughness. This of course would not be the case for a detuned fifth, such as 200 Hz and 305 Hz, since now the third partial of 200 Hz (600 Hz) clashes with the second partial of 305 Hz (610 Hz).

An extreme case of consonance (harmony) is the octave (e.g., 200 Hz and 400 Hz), where all the partials of the higher tone (400, 800, 1200 Hz) are already present as partials of the lower tone (200, **400**, 600, **800** ...), so the higher tone has nothing new to contribute. This then explains why two tones an octave apart sound so alike and in music are given the same name. The tone of 440 Hz and the octave 880 Hz are both called A but are distinguished by a subscript: A at 440 Hz is called A_4, while the octave above, A at 880 Hz, is called A_5. In this way, A_0 is near the bottom of the audible range.

To repeat: Tones with simple number ratios of frequencies are consonant because of absence of beats or roughness between their higher partials. We said "simple number ratios," but how simple do they need to be? An experiment shows that 6:5 is perceived as consonant (in earlier times it was considered dissonant!), while already 9/8 is not and 16:15 is very dissonant!

20.3 The Major Triad

A "triad" is a set of three tones often played together as a "chord." If you learned to sing or to play an instrument, you will know the names of the tones of the (Western) musical scale, like do–re–mi–fa–so–la, and so on, or C–D–E–F–G, and so on.

If you skip every other note starting on C, you get a sequence C–E–G, which is called the ***major triad***. What is special about it is the simple relationship between the three frequencies:

Frequency ratios of the major triad:	4	5	6
divide by 4	1	$\dfrac{5}{4}$	$\dfrac{3}{2}$

To get the frequency in Hz, multiply the three numbers by any number you like. For instance,

Multiply the sequence 4, 5, 6 by 50:	**200 Hz**	**250 Hz**	**300 Hz**
The tones in the usual notation are:	**C**	**E**	**G**

In reality, C is tuned to 264 Hz, but to keep the numbers simple, we use 200 Hz for the first tone. Remember, all that matters to music is the ratio of frequencies, not the frequency itself. If you used 264 Hz for C, E becomes $264 \times (5/4) = 330$ Hz. The first interval of the major triad, 5/4, is what we called a *major third*. The second interval, 6/5, is a *minor third*. The overall interval (C to G), 6/4 = 3/2 is a *fifth*.

What is special about the major triad is that it contains three simple-number ratios and thus sounds especially pleasing when the three tones are played either at the same time or in sequence.

20.4 Constructing a Scale: The Just Scale

Because people like harmony, a scale should permit a lot of harmonious intervals. With this idea in mind, we can construct a scale step by step, as described here. Through the physics of sound, therefore, we can explain how the currently used scale most likely came about.

To play music, we need more than just three tones; we need a "scale" with more tones. If C is the beginning of the scale, then the octave above is C again, but what frequencies do we put in between? Greek philosophers placed great value on numbers, and since they knew from experiments with vibrating strings that a fifth (3/2 ratio) is a perfect harmony, one idea was to keep going up in frequency in fifths—that is, repeated multiplication by 3/2. Already the second fifth goes beyond the octave, since $(3/2) \times (3/2) = 9/4$ is larger than 2. But they could go down an octave, which meant to multiply by 1/2, which makes the new frequency 9/8. This is nice when played after C, as in the very beginning of the tune "Are you sleeping ..." The scale based on fifths is called the Pythagorean scale. You can calculate the ratios for this scale yourself with a calculator or a spreadsheet on a computer. But the fly in the ointment is that multiplication by 3/2 (going up in fifths) followed by multiplication by 1/2 (going down in octaves) never makes a frequency that is double of the first frequency, so we cannot even make a perfect octave. What a shame!

Here we present another way to make a scale. Since the major triad is a perfect harmony, we build the scale by **adding more tones forming major triads**. Before we write down any numbers, here is the plan. The end result will be a sequence of frequencies for the tones of the scale:

C **D** *E* **F** *G* **A** **B** *C′* **D′** **E′**

Let us start with the triad we already know, called the major triad C–E–G, and add the octave of C (we can call it **C'** to distinguish it from the lower C). The plan is to link two further triads to the first one:

One triad starting on G: in musical notation, these will be tones called G–B–D', and one triad ending on C': in musical notation, these will be tones F–A–C'.

In other words, we will choose the frequencies such that when we skip every other tone of the scale we get perfect triads.

These additional two triads are reported in green and in blue:

C D E F *G* **A** *B* **C'** *D'* **E'**

C D E *F* **G A B** *C'* *D'* *E'*

Now we need to calculate the required frequencies. Again, we call the first frequency C = 1.

Figure 20.1 shows the three triads separately. To get the numbers for the green triad, starting on G, since we already fixed G at G = 3/2, we now must multiply by (5/4) to get the second tone: B = (3/2) × (5/4) = 15/8. The third tone must be (3/2) above the first—thus, D' = (3/2) × (3/2) = 9/4. The D an octave below will be (1/2) × (9/4) = 9/8.

To calculate the frequencies of the last triad, we must end the triad on C' = 2. So the first tone, a fifth below, must have a frequency 2 × (2/3) = 4/3 (remember, to go down a fifth, we divide by 3/2 or multiply by the reciprocal, 2/3). Finally, from F = 4/3 we go up a major third (5/4) and get A = (5/4) × (4/3) = 20/12 = 5/3.

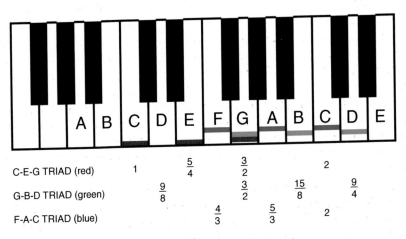

C-E-G TRIAD (red)	1		$\frac{5}{4}$		$\frac{3}{2}$		2		
G-B-D TRIAD (green)		$\frac{9}{8}$			$\frac{3}{2}$		$\frac{15}{8}$		$\frac{9}{4}$
F-A-C TRIAD (blue)				$\frac{4}{3}$		$\frac{5}{3}$		2	

FIGURE 20.1 ▶ The three major triads of the just scale are indicated on a piano keyboard.

Here is the final result of our labors. We have constructed a scale from three beautiful, perfect triads. This sequence of frequency ratios relative to the first tone of frequency 1 is called the *just major scale*. If the first tone of the scale is chosen to be C, the scale is called the just C-major scale.

To find actual frequencies in Hz, again you would need to multiply all those ratios by whatever number of Hz the first tone has. In the table below we called the first tone C, but the first tone of the scale need not be tuned to the actual C of the piano. You could start the scale on a lower or higher pitch.

Name		DO	RE	MI	FA	SO	LA	TI	DO
		C_4	D	E	F	G	A	B	C_5
Ratio to C_4		1	9/8	5/4	4/3	3/2	5/3	15/8	2
Decimal equivalent		1	1.125	1.25	1.33	1.5	1.66	1.875	2

Here we have used the notation C_4, C_5 for tones an octave apart. The preceding chart is valid for any other octave.

Extension to lower and higher octaves: Frequencies an octave above those shown in the chart are twice as large; frequencies two octaves above are $2 \times 2 = 4$-times as large; and so on. Frequencies an octave below are half the frequency, and so on.

How do we use these numbers?

Examples:

1. If C_4 has a frequency of 264 Hz, what is the frequency of F_4?

Answer: The ratio between F_4 and C_4 is 4/3.

$$\frac{4}{3} \times 264\,\text{Hz} = 352\,\text{Hz}.$$

2. If A_4 is 440 Hz, what is the frequency of E_4?

Answer: We need to find the frequency ratio between the two tones. Let us calculate the interval E–A:

$$\frac{E}{A} = \frac{5/4}{5/3} = \frac{5}{4} \times \frac{3}{5} = \frac{3}{4}, \text{ thus, } E_4 = (3/4)A_4 = (3/4)440\,\text{Hz} = 330\,\text{Hz}.$$

3. If G_4 has a frequency of 396 Hz, what is the frequency of D_5?

Answer: Calculate the interval $G_4 - D_5$. To find the frequency of D_5, an octave above D_4, we take the frequency of D_4 and multiply it by 2: $(9/8) \times 2 = (9/4)$.

The interval $G_4 - D_5$ is

$$\frac{D_5}{G_4} = \frac{9/4}{3/2} = \frac{9}{4} \times \frac{2}{3} = \frac{3}{2}, \text{ thus } D_5 = 396 \times \frac{3}{2} = 594 \, \text{Hz}$$

20.5   Whole and Half Tone Intervals

You will certainly be curious to know whether climbing up the just scale is like climbing up a ladder in which the rungs are equally spaced. Are all the intervals in this scale the same?

This is easy to figure out if you remember that an interval is the ratio (rather than the difference) between two frequencies. So you can simply take a calculator and find the ratio between adjacent numbers, always taking the higher number divided by the lower. In Table 20.1, we copy the just frequency table and underneath show the intervals.

If you want to get the answer as a fraction rather than decimals, you need to remember how to divide fractions: You multiply by the reciprocal. For instance, from F to G, the interval is

$$\frac{G}{F} = \frac{3/2}{4/3} = \frac{3}{2} \times \frac{3}{4} = \frac{9}{8}$$

We note two interesting results: The intervals are not at all the same but rather they come as two kinds: big intervals (called **whole-tone intervals**) and small intervals (called **half-tone** or **semitone intervals**). On the last line we label them as W and H (for whole and half, respectively). In Table 20.1 we also listed the frequency change from one tone to the next in percent change. For the whole-tone intervals the changes are 12.5% and 11.1%, while for the half-tone interval the change is 6.7%, nearly half of the larger intervals. So

TABLE 20.1 ▶ Increments Between Adjacent Tones of the Major Scale. W is Called a Whole Tone Interval, or Simply "a tone", H is Called a Half-tone or Semitone

Name:	C	D	E	F	G	A	B	C
Ratio to C:	1	9/8	5/4	4/3	3/2	5/3	15/8	2
Interval		9/8	10/9	16/15	9/8	10/9	9/8	16/15
in decimals		1.125	1.111	1.067	1.125	1.111	1.125	1.067
% change		12.5	11.1	6.7	12.5	11.1	12.5	6.7
Whole tone **W**, half tone **H**		**W**	**W**	**H**	**W**	**W**	**W**	**H**

we may call the whole- and half-tones 1 and 1/2. Any sequence of tones that has this particular pattern of intervals:

$$\text{Major scale intervals:} \quad \boxed{1 \quad 1 \quad \tfrac{1}{2} \quad 1 \quad 1 \quad 1 \quad \tfrac{1}{2}}$$

is called a *major scale*. Depending on the tuning of the instrument, the frequencies may differ a little bit from the just tuning, what is important here is that there are two kinds of intervals, large and small. Scales using two kinds of intervals, large and small, are called *diatonic scales*.

When you read in the concert program that a piece is in F-major, the composition uses a scale with a key note of F (the starting point of the scale) and again uses the preceding sequence of whole and half-tone intervals, starting on F. Major scales are widely used in Western cultures, but less so in African or Asian music.

This is remarkable! If you compare the preceding sequence of whole and half-tones to a piano keyboard, you now suddenly realize why the black keys are distributed in the familiar pattern: The groups of two black keys and three black keys, separated by a gap, correspond exactly to the groups of two whole tones and three whole tones, separated by a half-tone gap. The black keys are where there is a large interval, W. The black keys provide a half-step where there is a big interval. We need no black key where there already is a half-tone interval because there already is a small interval at this place.

It is important to remember the sequence of whole and half tone intervals, but we now see that the sequence is easy if you *picture the piano keyboard. Where the black key is missing, two adjacent white keys are a half-tone apart.*

The musicians who developed the scale centuries ago may not have thought about it this way and presumably were guided by musical taste rather than numbers, but it is certainly not by chance that our tuning using triads quite naturally led to two kinds of intervals, half-tone and whole-tone, distributed in this particular way.

20.6 Names of Intervals

Why is the interval from C to G called a "fifth"? Well, if you count all keys when you ascend the scale from C to G on a piano keyboard, indeed you find five, provided you include in your count the first key C as well. But it is somewhat strange to say that G is "a fifth above C," because it really is the *fourth* tone *above* C.

Having learned this quirk of musical nomenclature, it is now easy to deduce the names of other intervals and to find the frequency ratio for various intervals if the keyboard is tuned according to just tuning. Table 20.2

TABLE 20.2 ▶ Names of Some Intervals, Number of Semitones and Just Frequency Ratios

Name of Interval	Number of Semitones	Examples (Blue if Just)	Frequency Ratio for Just Intervals
Minor third	3	E–G, D–F	6/5
Major third	4	C–E, F–A	5/4
Fourth	5	C–F, D–G	4/3
Fifth	7	C–G, D–A	3/2
Sixth	9	C–A, D–B	5/3
Octave	12	C–C, D–D	2

lists the names of intervals and their frequency ratio. We also count how many "semitones" the interval contains, where we count whole tones as two half tones (also called semitones). *The content of this table is not to be memorized. Just look at the piano keyboard and simply count the semitones!*

Remember, when we constructed the scale, we started with a "major triad," C–E–G. The interval C–E is a third, and so is the interval E–G. But if you look at the keyboard or at the last row of Table 20.1, you may note that from C to E we go up four semitones (two whole tones), but for E–G only three semitones. A third consisting of four semitones is called a ***major third***, and a third consisting of three semitones is called a ***minor third***. The same idea is used for other intervals: C–D (1 whole tone = 2 semitones) is a major second; E–F (one semitone interval) is a minor second.

Table 20.2 shows that the just scale contains many intervals (ratios) that are consonant—simple number ratios. We calculate the ratio, for instance, for the minor third by looking up the frequencies for G and E in the just frequency table and dividing the higher frequency G = 3/2 by the lower one E = 5/4.

The adjective **just** ***is added to the name of a frequency interval if the interval has an EXACT simple number ratio such as 3/2.*** It is important to note that in "just" tuning not all intervals are "just"! Take the interval from D to A. It is a fifth and consists of seven semitones. But is it "just"? If we calculate the ratio of A = 5/3 to D = 9/8, we find—surprise!—40/27, and not 3/2. More about this later, but remember that in just tuning, only those intervals that have nice, simple number ratios and thus perfect harmony are called "just." For octaves one does not distinguish between just and nonjust because octaves are assumed to have a perfect 2:1 ratio no matter what tuning is used.

The third column of Table 20.2 shows several examples of intervals that are not "just." Check it out yourself by using the numbers from the just frequency table.

Example: A 50 cm long string of a viola is tuned to G_4. How far from the nut must one place the finger on the fingerboard to play C_5? What is the name of the interval? (The nut is the far end of the fingerboard, right before the tuning pegs.)

Answer: Counting keys on the piano keyboard, the interval is a fourth. Check to see if it is a "just" fourth: The ratio is 2 divided by $3/2 = 2 \times (2/3) = 4/3$. It is a just fourth (Table 20.2). The new length of the string must be shorter than 50 cm in the 3/4 ratio. The new vibrating length is 50 cm $\times$ (3/4). The distance from the nut is what remains, 50 cm $\times$ (1/4) = 12.5 cm.

For completeness let us mention another problem of naming intervals. In Table 20.2 we said that a fourth has five semitones. If you try this out on the picture of the piano keyboard by placing fingers on C and F and moving the pair of fingers up the scale, you find that indeed the fourth always includes one half-tone interval—with one exception! From F to B is indeed a fourth but an exceptional one, having three whole tones or six semitones instead of five. It is called the ***augmented fourth***.

No such exception arises for the fifth, as you can easily convince yourself. But for the sixth you note that usually it spans one semitone and four whole tones, nine semitones altogether, as shown in Table 20.2. There is one exception: From E to C is indeed a sixth, but this one is the exception because it spans eight semitones instead of nine. It is thus a smaller interval than the normal sixth and is called the ***minor sixth***.

20.7   Transposing: Why Black Keys?

If you look at the intervals of the "Frère Jacques" song at the beginning of this chapter, you will find that three tones correspond to **C–D–E** of the C-major scale—namely, frequencies (with respect to C = 1) of 1, 9/8, and 5/4. Now if the pianist plays the song in C-major, using the white keys of the piano, the singer may find it a bit too high in pitch for his voice. He may ask the pianist to play her part "in a lower key"—starting a tone lower on B. The song needs to be "transposed" to a lower key.

When "transposing" we need to keep the same musical intervals, which means the same number of semitones between the various tones. Let us attempt to transpose a C-major scale down to B-major. The sequence of whole and half-tones: 1, 1, 1/2, 1, 1, 1, 1/2 must be maintained, but we need to begin on B_3.

If we use the white keys of the piano, we are in trouble because playing B–C–D instead of C–D–E is not what we want; it sounds not just lower but all wrong! C–D–E has two whole tone intervals in sequence, but B–C–D starts out with a semitone instead. The solution is to make a new tone, one whole tone above B. This tone will be a half tone above C. We use the black key to the right of C, tune it accordingly, and call it $C^\#$ (called C-sharp). How do we tune $C^\#$? It must be a 9/8 ratio to B because we want to maintain the original interval. What's the next tone after $C^\#$? D would not be right because it is only a half-tone above $C^\#$, so we use another black key, the one to the right of D and call it $D^\#$.

By now you should get the idea: To transpose a major scale, we keep the sequence of whole and semitones, 1, 1, 1/2, 1, 1, 1, 1/2, but start on a different tone and we use the black keys as needed to get the desired whole and half-tone steps. If you continue, you will find that the sequence of tones for the B-major scale is

$$B \ C^\# \ D^\# \ E \ F^\# G^\# A^\# B.$$

Note that we did not need to calculate any frequencies. Instead, we just sketch the piano keyboard and count the number of whole tones and semitones. Music students have to memorize which black keys (also called "accidentals") to use for each scale. Here instead, we learned to reason it out! A tone is called "sharp" or "flat," depending on whether it is the black key on the right or the left of the corresponding white key. So which of the names do you use? The rule is that you use the same letter only once. In the preceding example, we would not replace $D^\#$ by E^b (E-flat).

20.8 Perfection Sacrificed: The Tempered Scale

We talked about transposing and why there are black keys on the piano. Transposing from C-major to B-major required many black keys. Does the addition of the black keys take care of everything? There is a problem we swept under the rug thus far. We called the interval from C to D a whole tone, and from D to E a whole tone also. But these are not really quite the same. Look back to the chart of intervals in the Just Scale (Table 20.1). If we tune the piano to the just scale and want to transpose from C-major to D-major, we are in trouble immediately because the first interval of the major scale is supposed to have a 9/8 ratio. But if we start playing the tune on D, the first interval is 10/9 rather than 9/8. Maybe this is not a big problem because we usually don't play two adjacent tones at the same time anyway, so why worry about a little mistuning? But worse is to come: remember the beautiful, perfect, 3:2 ratio, the fifth? In just C-major tuning, C to G was a perfect 3:2 ratio, but if we start our tune on D and play the tone a fifth above (A), the interval will be an ugly, out-of-tune fifth—as we see if we work out the frequency ratio between D and A.

To figure it out, take the higher frequency (A) from the just frequencies in
Table 20.1 divided by the lower frequency (D):

$$\frac{A}{D} = \frac{5/3}{9/0} = \frac{5}{3} \times \frac{8}{9} = \frac{40}{27} = 1.482$$

The fifth from A to D in just tuning is far from a perfect fifth. Harmony
depends on coincident overtones. Here, if we make A = 440 Hz, the frequency
of D will be 440/1.482 = 296.9 Hz. Harmony requires that the higher partials
coincide. Here, the second harmonic of the A is 880 Hz, but the third harmonic
of the D is 890 Hz. The 10 Hz difference produces a 10 Hz beat frequency,
which we perceive not so much as rapid beats but as unpleasant roughness.
For the same interval one octave or two octaves lower, all the frequencies
would be half or a quarter as much, which means now the 10 Hz difference
becomes 5 Hz or 2.5 Hz, producing very recognizable beats.

A very convenient way to picture this is to mark the frequencies of the
scale on a logarithmic number line. To review the main point: On a log scale,
the numbers are arranged in such a way that equal ratios (or equal factors)
are represented by equal distances. This is perfect here because equal ratios
of frequencies means equal musical intervals so that on a log scale equal
intervals are shown as equal distances. This makes it easy to see what goes
wrong in transposing. *Transposing* means to shift the entire scale up or down.
Figure 20.2 showed the just scale marked on a log scale (green marks) and
underneath, the same set of marks shifted to the right so that the first mark
has moved from C to D. One notes that to play a scale starting on D, some
of the green (C-major) frequencies are off. In particular, the (green) E and A
are too low in frequency; they are flat, while G and B are perfectly fine. This
is confirmed by the calculation we just did for the interval from D to A. You
might want to do the same for other intervals.

Help in visualizing scales:

• Equal musical intervals—equal frequency ratio
• On a "multiplicative" number line (= log scale), equal ratios are
  equidistant

Advantage: in Figure 20.2, equal intervals have the same length.

We conclude from this that no matter how we tune the two black keys, we
still have to retune the white keys when changing from C-major to D-major.
Now we have two choices: Either we stay with C-major and keep out of trouble or we retune the keyboard every half hour when we want to play a different
key! To stay with C-major limits the musical expression. Although we said
that the tune: "sounds the same", when it is transposed to D-major or E-major

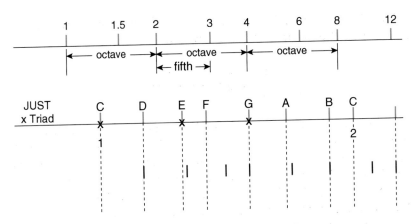

FIGURE 20.2 ► A logarithmic scale is used to explain the problem that arises in transposition when just tuning is used. On the line labeled "Just" the frequencies of the just scale are marked (green marks). If we shift the scale up one tone (transposing from C to D), one obtains the short black marks below, as you can check yourself by copying the green scale and comparing to the black scale; they are the same but shifted. The short black marks indicate the frequencies we would need for a just scale in D. We see that G and B are tuned properly, but E and A would have to be retuned. The black markers between F and G and between C and D show to what frequency the black keys F# and C# need to be tuned.

(and so on), a higher key makes it sound different just because it is higher, brighter, and more sparkling.

Until the 18th century, keyboard instruments were retuned before switching—for example, from C-major to D-major. The black keys, of course, one can tune prior to the performance, but the white keys needed to be retuned. Depending on what key one wants to use, in some cases only a few tones are off and need adjustment. In Bach's time, music was played in very few keys and the keys were chosen so that between pieces only a few notes of the keyboard (harpsichord) had to be adjusted. But still, it was a great nuisance. Moreover, church music that included pipe organs was the driving force behind much of the musical creativity of great composers, among them Frescobaldi and J. S. Bach. Retuning organ pipes was and is completely impractical.

Composers and musicians have experimented with alternative tunings to alleviate these problems, and some of them were rather successful, at least for applications limited to specific keys, such as Bach's version of tempered tuning. It is too complicated to describe these tuning compromises here, so we proceed to describe the contemporary radical solution.

Modern tunings use a strikingly simple scale—that is, a simple sequence of tone frequencies. Remember, the troubles started because in just tuning, the whole tones are not all alike, some have 9/8, others have 10/9 frequency ratios. And in addition the semitone was not really half of a whole tone. That's why we had difficulty transposing from C-major to D-major. The idea of the modern *tempered tuning* is to *divide the octave into 12 identical semitone intervals*. Each whole tone then is exactly two semitones, and all are alike. Why 12 intervals? Because we want to retain as much of the just tuning as possible, and in just tuning each octave had five whole-tone intervals and two semitone intervals, which combined give $5 \times 2 + 2 = 12$ increments.

What would the frequency ratio have to be, then, for a whole tone and for a semitone? Remember that "same interval" means same frequency ratio. Each time we go up a semitone, we multiply the previous frequency by some number x, using the same number x 12 times so that all intervals are the same. But after 12 of these steps, we must have the octave of the first tone. Expressed mathematically:

$$x \cdot x \cdot x \ldots\ldots\ldots \cdot x = x^{12} = 2$$

So now we need to figure out what number multiplied by itself 12 times makes 2!

The answer is:

$$x = \sqrt[12]{2} = 1.0595\ldots$$

This is the semitone ratio of the tempered scale.

> **Tempered semitone frequency ratio: $x = 1.0595\ldots$**

In the tempered scale the frequency ratio of any two adjacent keys of the piano (black or white) is the same and equals $1.059\ldots = \sqrt[12]{2}$. The advantage of the tempered scale is that there is no difficulty transposing because the intervals between adjacent (black or white) keys on the piano are all exactly identical.

This only says that there is no loss when transposing without any need to retune, but it does not say anything about how badly out of tune the scale became, due to the radical change to equal semitone steps!

To see what happened, we calculate the tempered ratio for various important intervals and compare to the desirable just ratios, as reported in Table 20.3.

The tempered whole-tone ratio is $x^2 = 1.1225$. This number is in between the two whole-tone ratios of the just scale. For the other intervals, the tempered ratios are easily calculated by multiplying x by itself a number of times equal to the number of semitones in the interval (or use the x^y key on your calculator).

TABLE 20.3 ▶ Comparison of Tempered and Just Intervals

Name of Interval	Number of Semitones	Tempered Ratio	Frequency Ratio for Just Interval	Deviation %
Minor third	3	1.189	6/5 = 1.200	0.9%—bad
Major third	4	1.260	5/4 = 1.250	0.8%—bad
Fourth	5	1.335	4/3 = 1.333	0.12%—good
Fifth	7	1.498	3/2 = 1.500	0.07%—very good
Sixth	9	1.682	5/3 = 1.666	0.9%—bad
Octave	12	2	2	

A deviation of 0.1% is still harmonious while one that approaches 1% used to be considered dissonant. These are labeled "Good" and "Bad", respectively.

Table 20.3 shows that the tempered intervals are not far from the just intervals. And most importantly, the strength of the tempered tuning is that the tuning is equally good (or equally bad!) for all keys—that is, one can transpose at will without any further loss in harmony and no need for retuning. The **weak points** of the tempered scale are the relatively large deviations for the **major and minor third**, which are off nearly 1 percent, or 1 Hz for every hundred Hz in frequency. This means that for tones near 200 Hz, the tempered thirds are off about 2 Hz—enough to cause significant imperfection in the harmony. These tempered thirds were at one time considered objectionable, particularly because Baroque composers frequently used parallel lines spaced in thirds, but modern audiences find them very acceptable. On the contrary, we are now so used to the tempered thirds that some time ago, when a music critic listened to a classical concert in just tuning (without being aware of it), he commented in a newspaper review that the performance was out of tune.

In Table 20.3 note that the tempered fifth, the most important of all consonant intervals other than the octave, deviates by less than 0.1% from the just ratio. Also, the fourth is very satisfactory. In modern music terminology, the tempered fourth and fifth is referred to as a "perfect" interval, even though it is not mathematically perfect.

A table of frequencies for tempered tuning with $A_4 = 440$ Hz is given in Table 20.4. A scale that, as the tempered scale, assigns a definite frequency (or frequency ratio) to all 12 semitones (white and black keys) is called a **chromatic scale**. For comparison, the frequencies of the just C-major scale are also listed in the table. The just scale is not a chromatic scale because it does not assign a unique frequency to the black keys; they are tuned differently as needed when transposing.

TABLE 20.4 ▶ Just and Tempered Scale: The Table Lists Frequency Ratios (with respect to C$_4$) and Frequencies for the Middle Octave of the Piano (assuming A$_4$ = 440 Hz for both scales)

	Just		Tempered	
	Ratio	Frequency	Ratio	Frequency
C$_4$	1	264	1	261.6
C$_4^\#$ D$_4^b$			1.059	277.2
D$_4$	9/8 = 1.125	297	1.122	293.5
D$_4^\#$ E$_4^b$			1.189	311.1
E$_4$	5/4 = 1.25	330	1.260	329.6
F$_4$	4/3 = 1.333...	352	1.335	349.2
F$_4^\#$ G$_4^b$			1.414	370.0
G$_4$	3/2 = 1.5	396	1.498	392.0
G$_4^\#$ A$_4^b$			1.587	415.3
A$_4$	5/3 = 1.666...	440	1.682	440
A$_4^\#$ B$_4^b$			1.782	466.2
B$_4$	15/8 = 1.875	495	1.88	493.9
C$_5$	2	528	2	523.2

String orchestras or string quartets are not restricted to particular intervals. Books sometimes claim that good professional string quartets play the just scale for perfect harmony. Frequency measurements on first-rate professional string quartets show that they deviate from the just intonation and in fact play somewhere between the just and the tempered tuning.

Many musicians and music theorists are still searching for better ways to tune fixed-frequency instruments like the pipe organ. Not everyone accepts that the modern tempered tuning is the best solution. Particularly, lovers of Baroque music often use tunings other than the tempered tuning, and some musicians have made a name for themselves for their performances and recordings using historic Baroque tunings.

Example: A guitar string is 60 cm long. The frets on the guitar are placed according to tempered tuning with a semitone from one fret to the next. What must be the distance from the nut (end of guitar) to the first fret?

Answer: The string must be shortened in the ratio 1:1.0595 (1 semitone ratio). The shortened string must be 60 cm/1.0595 long = 56.6 cm. This is 60 − 56.6 = 3.4 cm shorter than the original string. The first fret must be 3.4 cm from the nut.

20.9 Major and Minor Scales

As explained in section 20.5, if you remember the layout of the white and black keys of the piano keyboard, you can reconstruct the sequence of whole and half tones.

Intervals of a major scale: $1 \quad 1 \quad \dfrac{1}{2} \quad 1 \quad 1 \quad 1 \quad \dfrac{1}{2}$

As you might imagine, other sequences are possible. The next common choice results in the minor scale.

Intervals of a minor scale: $1 \quad \dfrac{1}{2} \quad 1 \quad 1 \quad \dfrac{1}{2} \quad 1 \quad 1$

This is the sequence that occurs if you play the white keys of the piano starting on *A*. Musicians usually modify this scale because the whole tones at the end of the scale sound strange, but we need not go into this here. If you play a minor scale, it sounds very different from the major scale, even though it is the same sequence of whole and semitone intervals. We just started the sequence at a different place. Compositions in minor keys tend to sound more subdued, somber, and even sad.

You can, of course, create new scales by starting the sequence in different places, such as D or E. These scales were used in Greek music and early Christian church music (called Gregorian modes). These go under the names of Dorian, Lydian, and Aeonian modes. Note that when we use the term *modes* to denote a particular sequence of whole tones and semitones (such as major mode, minor mode, Greek modes), this denomination has absolutely nothing to do with sinusoidal motion of string and pipe modes.

Some modern composers use no particular sequence of whole and semi-tones, but instead they use all semitone intervals within the octave (that is, all 12 white *and* black keys within the octave are considered equal). This is referred to as *12-tone music*. Remember that the intervals of the just scale came up with large and small intervals—that is, with whole and half steps, in order to produce harmonious intervals. In 12-tone music, consonance (harmony) plays no important role. Some people like it nevertheless.

20.10 The Natural Scale

A sequence of frequencies that plays a special role in music is the so-called *natural scale*, also called the *harmonic series*. It is the scale that is played on a "natural horn"—that is, a pipe without valves or tone holes (finger holes), playing only **multiples of the fundamental** frequency. Here, we simply label the frequencies 1, 2, 3, and so on, with the understanding that to obtain the frequencies in Hz, we multiply all of them by the fundamental frequency in Hz.

Natural scale 1 2 3 4 5 6 7 8 9 10 11 12
               C C G C E G X C D E   X   G

Now we want to know how this scale relates to the just scale, which is also made up of whole numbers. You may recognize right away that the major triad (like C–E–G) is part of the natural scale. If the "natural instrument" is tuned to C, we can label the tones of the natural scale as shown: the first tone is C, and so are the tones one, two, and three octaves above (2, 4, 8). The sixth tone of the natural scale is a fifth (3/2 ratio) above the fourth tone (C), so it is G. If the sixth tone is a G, so are all other tones an octave below or above (3, 12). Bugle tunes use tones of the natural scale. What the tones are if the bugle is tuned to G rather than C you will be able to figure out by extending the frequency table for the just scale to other octaves. The seventh and eleventh tones of the natural scale are marked X because they do not exactly match tones of the just scale.

20.11 Questions

Use the table of frequency ratios given in the text.

1. (a) If you tune A_4 to 440 Hz, what is the frequency in Hz of D_5 in just tuning? (Hint: Use the table of just frequency ratios given in Figure 20.1.)
    (b) What must be the frequency in Hz of F_4 in just tuning?
    (c) What is the frequency of a tone a just minor third above A_2?

2. The D-string of a guitar is tuned to D_3. The string is 60 cm long. Where must you place the fret to play the octave above D_3? To play E_3? (Remember: guitars use tempered tuning.)

3. An instrument plays the "natural scale." It the first tone of this scale is 200 Hz, what are the frequencies of the following four tones?

4. An open pipe is tuned to G_3. What tones on the musical scale are the next four modes of the pipe?

5. What is the fundamental frequency and the length of the preceding pipe?

6. A French horn is 350 cm long.
    (a) What length of tubing should be added to decrease the frequency by a tempered semitone?
    (b) By two semitones?
    (c) By three semitones?

7. How many semitones are there in the following intervals: octave, fifth, fourth, major third, minor third?

▶ Chapter 21

Musical Instruments

21.1 Structure of Musical Instruments

Musical instruments are intended to produce sounds of a particular pitch and amplitude. In most cases, these sounds are more or less periodic over many oscillations and usually involve a complex wave shape that contains a number of partials, often 20 or more.

The physical structure of an instrument can be represented by the diagram in Figure 21.1. At the center of the instrument is an oscillating system (here called a *resonator*), such as the string of a violin, the air column of a trumpet, or the drumhead of a kettle drum. To set the oscillator into motion, we need *an excitation device*, such as the bow of the violin, the lips of the trumpet player, or the piano hammer hitting the string. The excitation device in turn requires *a source of energy*—for instance, the bowing arm of the string player, the lung of the flutist, or the compressor that blows air into the organ pipe. Finally, the energy supplied by the energy source to the oscillator is given off as a sound wave, which requires some sort of a *radiator*.

For some instruments, the radiator is clearly visible to the eye. The sound board of a piano, the body of the guitar, or the violin. The string itself is a very poor radiator because it is much thinner than the wavelength of the sound in air being radiated. Many brass instruments and woodwind instruments have a flared end (a "bell" or a "horn") to improve sound radiation. The sound radiation is usually stronger in particular directions, for the violoncello roughly perpendicular to the body of the instrument, for brass instruments primarily along the axis of the bell. Composers are aware of these technical details. For

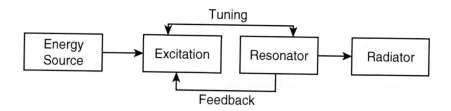

FIGURE 21.1 ► Schematic of the physical structure of a musical instrument.

instance, Bruckner specified in a loud trumpet passage that the trumpet is to be played "with raised bell." The flute does not have a bell, but rather the radiator for the flute is the air escaping from the open ends and the finger holes, none of which are very efficient in radiating sound. Indeed, the flute in the orchestra struggles to be heard unless the composer gives it a short solo part now and then.

21.2 Excitation Mechanism

For percussion instruments or plucked strings, the excitation is relatively simple: the oscillator is hit once, and the excitation consists of a momentary impact. The excitation of a **steady tone** is more difficult: the tone sometimes must last for several seconds, which means that the resonator will make many hundreds of oscillations during a given tone. The musician has therefore a seemingly impossible task: he must control the excitation so that it is at resonance with the oscillator. To put it in simple words, suppose the trumpet plays an 880 Hz A_5 for three seconds. The air column in the trumpet during this time makes about 2600 oscillations. This then requires that the air column (the resonator) be pushed by the musician 2600 times but with air puffs that are in step (in resonance) with the oscillator. By what magic does the musician accomplish this task?

The answer is that the trumpet player presses his lips against the mouthpiece and blows to make a buzzing vibration of the lips. You can buzz the lips yourself even without a mouthpiece by pressing the lips together fairly hard and blowing. When the musician first starts blowing into the trumpet, the pressure variation from the buzzing lips travels along the trumpet and is reflected again and again. It is the pressure wave reflected from the end of the trumpet that locks the lip frequency to the pipe frequency, as long as the musician learned to buzz somewhere *near* the oscillator frequency. In effect, the excitation (buzzing lips) is the driving force on the resonator, but the resonator in turn affects the driving force (the lips) to keep it at the resonance frequency. This process is described as *feedback between oscillator and excitation.* Somewhat the same

FIGURE 21.2 ▶ Air flows through a narrow slot rapidly enough to make turbulent eddies in alternation up and down. The periodic pressure variations set the air in the pipe into oscillation. The pressure wave reflected from the end of the instrument in turn affects the eddies (feedback).

happens in organ pipes or a soprano recorder, in which air flowing through a slit is directed towards a sharp edge (Figure 21.2). The air flow is fast enough for the flow to make turbulence ("eddies" or vortices, the same you see in a fast flowing river behind a tree branch sticking out of the water). The oscillation of the turbulent air—the so-called "edge tone"—sets in motion the air in the pipe. But the reflected wave from the end of the pipe in turn deflects the air stream, so by feedback, the oscillations of the air stream are locked to one of the modes of the pipe. Here again, the frequency of the air oscillations by themselves must be close to one of the pipe modes. The recorder player can excite the second mode of the pipe by blowing harder and by starting the air stream very suddenly, because now the air stream oscillations tend to become faster and can be close to the second mode instead of the fundamental. The player "overblows" to the octave.

What we just described applies in a similar way to the flute, where the slit is formed by the player's lips and the edge is the sharp edge of the mouth hole. Almost everyone has at one time blown across the opening of an empty bottle to make it sound. Again, you create an edge tone that is close to the mode frequency of the bottle itself.

Reed instruments, including the clarinet, saxophone, oboe, and bassoon, have one or two strips of tapered, flexible pieces of cane or plastic (reeds) in the mouthpiece. Again, these form a narrow opening through which the player's air stream passes. The reeds vibrate against one another (double reed instruments, e.g. oboe) or against the fixed mouthpiece (single reed instruments, e.g. clarinet), interrupting the air stream. Again, the reed frequency is locked into one of the modes of the instrument. The player can choose which mode she wants to excite by changing the lip pressure on the reed and by changing the speed of air flow. In this way she can, with practice, overblow the instrument to make the pitch jump from one mode to the other. Without the ability to overblow, the range of pitch available to the players of wind instruments would be far too limited.

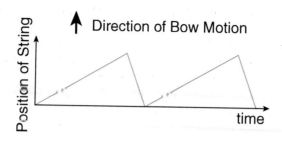

FIGURE 21.3 ▶   Schematic representation of a bowed string. The bow is made slightly sticky with rosin so that the string gets pulled aside when the bow is pulled across the string. At some point in time, the string snaps back and the process repeats. The pulse reflected by the end of the string helps to make the string shake loose from the sticky bow at the right moment.

For bowed strings, the musician pulls the bow across the string and sets the string into motion. Again, the driving force on the oscillator (the string) has to be in tune with the natural frequency of the string. How does the player wiggle the bow exactly 440 times every second when she bows the A-string? She doesn't! Again, feedback comes to the rescue. The string player rubs rosin on the bow to make it slightly sticky. When she pulls the bow across the string, the string is pulled aside (Figure 21.3) until the restoring force is too large and the string loosens from the bow and snaps back. This pulse travels down the string and is reflected after time T, the round-trip travel time on the string. The reflected pulse, when it arrives at the bow, helps to shake the string loose from the bow and the string snaps back at the right moment, so again the mode frequency of the oscillating system (string) feeds back in the excitation (bow) to keep it in tune. Again, the musician must learn to bow with the right pressure and speed of the bow to be near the desired frequency; otherwise the result is the awful scratchy sound of a beginner's attempt to bow a string.

21.3   Playing a Tune

To play a tune, the musician must be able to change the frequency of the sound. This requires changing the frequency of the oscillator to resonate at a sequence of frequencies, which correspond to the musical scale. This is a large and interesting topic, on which we make only a few brief comments.

1. **Strings:** Strings are tuned to the right frequency by changing the tension with the tuning pegs. Once tuned, the frequency is changed by pressing the string against the fingerboard. This shortens the vibrating length of

the string and increases the frequency in inverse proportion to the length of the string. All guitars have frets on their fingerboards. *The frets are placed according to tempered half-tone intervals*. Consequently, as you go up the fingerboard, frets get closer and closer because for each successive half-tone the string needs to be shortened by the same fraction of the remaining length. When playing the guitar, the finger is not placed *on* the fret but between frets so that the position of the fret accurately determines the length of the vibrating part of the string. Frets make it easy to play in tune (provided the strings were initially properly tuned with the tuning pegs) but have the disadvantage that they are not flexible. The frequencies are therefore fixed to tempered tuning, and one cannot produce slides or a vibrato.

In keyboard instruments, each tone has its own string, properly tuned to the "right" frequency. One interesting aspect of piano tuning is that the octaves are not precise 2:1 ratios! This is related to string stiffness (Chapter 17), which increases the frequencies of the higher modes of a string a little bit. When you play two tones an octave apart on the piano, such as A_4 and A_5, you certainly want the second partial of A_4 to be in perfect tune with the fundamental of A_5. But because of string stiffness, the second partial of A_4 is a little more than twice the fundamental. Thus, the A_5, tuned to the second partial of A_4, will be a little sharp, a little more than twice the A_4 fundamental; it will be a "stretched octave." For one octave, the effect is small, but by the time you keep doing this, one octave after the other, the errors accumulate. If one compares the lowest tone of a grand piano to the highest one about seven octaves above, there is nearly a half-tone discrepancy compared to the mathematical 2^7 ratio. One can of course now synthesize sounds at will with computers; one can even play in just tuning, but it appears that the slight imperfections of real instruments are preferred by musicians and listeners.

2. **Woodwinds:** The pitch of the woodwind instruments is controlled by a number of holes drilled in the body of the instrument, called tone holes or finger holes. The lowest tone of a woodwind, such as a flute, oboe, bassoon, or clarinet, is played by covering all the tone holes. To play the tones of the musical scale, the player opens holes along the body of the instrument in succession, starting from the bottom, to shorten the air column and consequently raise the pitch. Proper placement of the holes is not as simple as placing frets on the guitar because the pressure node in a wind instrument is not exactly at the first open hole. The holes simply cannot be made large enough to assure a pressure node. This is seen clearly if you close a hole at the bottom end of the instrument and keep an intermediate hole open. If the open hole

had the effect of making a pressure node, with air escaping freely as if the pipe were cut off at that point, then closing a finger hole at the far end of the instrument should not change the pitch at all, but it does! Therefore, placement of tone holes and design of fingering patterns to keep the instrument in tune over three octaves is a challenge. All modern woodwinds use tone holes too large to be covered with a finger. Mechanically linked, padded covers over the tone holes ("keys") allow large tone holes and free choice of location of the tone holes. This, coupled with the ability of the player to overblow the instrument to higher modes, allows about three octaves in half-tone intervals and pretty accurate tuning. It is known that Mozart only reluctantly composed for the flute on commission when he needed the money because the flutes of his time could not be played accurately in tune. If the modern flute (designed by Böhm based on physics calculations) had been available to Mozart, we probably would enjoy many more of his flute compositions.

3. **Brass Instruments:** Brass instruments include the trumpet, tuba, trombone, French horn, and so on. The air column in these instruments is never just a straight pipe, but rather brass tubing that is bent into beautiful curves, because otherwise the instrument would be too large to handle. The same can be said for woodwinds playing low frequency tones, such as the bassoon, which is 10 feet long. The sound wave thus travels along a curved trajectory. As long as the bends are smooth and have no abrupt corners that would reflect waves, a curved pipe works just as well as a straight one but is a lot easier to handle and to play.

   Brass instruments change the frequency of the resonator by adding lengths of tubing. The slide used in the trombone is a simple and very transparent method to change the length of the pipe. Again the player needs to overblow to higher modes. Without overblowing, the change in length of the air column needed to play three octaves would be impossible to achieve within the physical limits of a person's arm length.

   Trumpets have three valves to change the length of the pipe. Other brass instruments sometimes have four or even five valves. When the trumpet player depresses a valve, a short length of tubing is added to the trumpet. To lower the pitch a semitone (ratio 1.059), we need to multiply the length of the trumpet by 1.059—that is, we add to the trumpet a little loop whose length is 5.9% of the original length. A second valve lowers the pitch by two semitones and a third valve by three. So all three valves depressed should lower the pitch by six semitones. But if you do the math you find that this is not quite true, because with increasing length of the air column, the amount of space per semitone to be added gets larger. For this reason the trumpet is provided with a small slide operated by the thumb, which permits the musician to correct the pitch.

What comes after six semitones? The answer is seven, of course, but seven semitones is a fifth, and the trumpet player can produce a fifth by overblowing from the second mode ($2f$) to the third mode ($3f$).

The trumpet and the trombone are cylindrical pipes. They are simply tubes of a constant diameter to which a bell and a mouthpiece are added. The player makes a pressure antinode at the mouthpiece. A pressure antinode at one end is characteristic of a closed pipe, so we should *expect mode frequencies that are odd multiples* of the fundamental. But the higher modes of brass instruments are in fact $2f$, $3f$, $4f$, *as if they were open pipes*. The reason is that the higher modes of these closed pipes are shifted intentionally by proper design of the bell and the mouth piece to correspond to $2f$, $3f$, $4f$, and so on. The frequency f is not actually a resonance mode of the instrument and is referred to as a "fictitious" fundamental. Other brass instruments, e.g. French horn and tuba, are conical over most of their lengths.

Example: The air column of the French horn is 2.6 m long. A piece of tubing is added by depressing a valve of the instrument. How long must the added piece of tubing be to lower the pitch by a semitone?

Answer: The semitone ratio of frequencies is 1.0595. The new length must be 260 cm × 1.0595 = 275.5 cm. The additional length is 275.5 − 260 = 15.5 cm.

4. **Voice:** When you sing a steady tone, air passes through the narrow opening between your vocal folds, which sets them into oscillations similar to the oscillations of the reeds on a woodwind. The singer can control the frequency of these oscillations by muscles, which change the tension on the vocal folds (also called "vocal cords"). But the air pressure variations are far from sinusoidal; rather the opening and closing of the vocal folds makes a sequence of sharp, periodic puffs of air with a large number of Fourier components in the sound spectrum. Air in the throat and in the mouth ("oral cavity") is set into oscillation by these puffs of air. Once again, this is an oscillator set in motion by a driving force. As far as resonators go, the throat and oral cavity have a large amount of damping and thus a wide resonance curve. The only components of the vocal folds' sound spectrum that are re-enforced and thus strong are the ones that resonate in the throat or in the oral cavity. In Chapter 19 we saw a spectrum of Willy's voice. The vocal folds produce 220 puffs of air every second so that the partials are all multiples of 220 Hz. The

sound spectrum in Figure 19.7 shows that the resonance in the oral cavity reinforces some frequency regions of Willy's voice. *These are called the formants of the particular vowel. Singing a **different** vowel of the **same** frequency would have the **same** set of partials (multiples of 220 Hz) but **different formants**.*

5. **Percussion Instruments:** Drums, bells, xylophones, and so forth are made to oscillate when struck or hit by hands or sticks. Large drums are the most powerful orchestral instruments. A single drum can compete with an entire orchestra in sound power output. For drums, the oscillations are heavily damped and thus decay rather rapidly. A special property of percussion instruments is that the **higher partials are nonharmonic**. Since the tone does not last very long, nonharmonic partials are not perceived as unpleasant. Some Indian drums (the tabla) use drumheads loaded by adding sticky material near the center of the drum, which permits shifting of the higher partials to be nearly harmonic. For bells and cymbals, the tone may last for a second or more, but the higher partials decay quickly so that very soon only the lowest partials are heard. These are tuned to be nearly harmonic.

Drums and bells have modes like strings: motions in which all parts of the instrument move at the same frequency and therefore undergo simple harmonic oscillation. As an example, modes of a drumhead are shown in Figure 21.4, in which the dashed lines show the positions of nodes (nodal lines). Along one of the nodal lines, the drumhead does not move, as one can show by sprinkling sand or sesame seeds on the drum. On either side of a nodal line, the drumhead oscillates in opposite

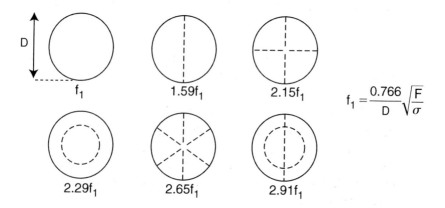

$$f_1 = \frac{0.766}{D}\sqrt{\frac{F}{\sigma}}$$

FIGURE 21.4 ▶ Modes of a drumhead. The higher partials are not multiples of the fundamental—that is, the partials are not "harmonics." The frequency formula is explained in the text.

directions, just like on a string the motion is opposite to the left and the right of a node. To excite a drumhead to a single mode, one can use a loudspeaker placed near the drumhead in a place where the desired mode has an antinode. The formula in Figure 21.4 shows similarity with the string formula. Here the quantity F is the pull per unit length of the circumference, σ is the mass per unit area of the drum head, and D is its diameter.

21.4   Questions

Note: The names of intervals remain the same when you go from just to tempered tuning. To find *names* of intervals, you need NOT calculate the frequency ratio but instead count the number of semitones. For instance, an interval of four semitones (two whole tones) is called a major third no matter what. Only if you want to know if it is a just interval do you need to check the actual frequency ratio, which would have to be 5/4 to be called a just major third.

1. The six strings of the guitar are tuned to E_2, A_2, D_3, G_3, B_3, E_4. Name the intervals between adjacent strings (NO calculation needed; instead look at keyboard and count semitones).

2. Are the whole tone intervals all the same in *just* tuning? In *tempered* tuning? Compare C–D and D–E.

3. A 60-cm-long guitar string is tuned to E_4. Where must I place a fret (i.e., how much must the string be shortened) to play a fourth above E? What is the tone called?

4. A viola string is 80 cm long and is tuned to G_3. What tone will the string sound when the musician reduces the oscillating length to 60 cm by pressing a finger on the fingerboard?
   (a) What is the length ratio?
   (b) What interval does this ratio represent?
   (c) What is the tone for 30 cm vibrating length?

5. Assume that the vibrating length of the air column in a brass instrument is 150 cm. Pressing a valve adds a certain length to the air column. How much length should be added to the air column to lower the pitch by one tempered semitone?

6. To blow A (440 Hz) on a trumpet, the air stream from the lips of the musician must vary 440 times a second. Explain how feedback works to keep the lips opening and closing at the right instance.

Chapter 22

Solutions to Problems

Chapter 2

1. **Virtual.** The rays reflected from a given point of my face (e.g., the tip of my nose) do *not* cross anywhere after reflection from the mirror, but rather they *seem* to come from a point behind the mirror.

2. The light beam is kinked toward the normal (see figure below).

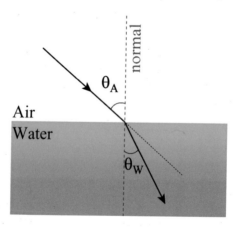

3. Light rays are always reversible. That is, you can use the same figure on the left, invert the light ray (bold arrow), and see that when going from water to air, the ray is bent away from the normal.

Physics in the Arts, Revised Edition

4. A light beam leaving the water perpendicular to the surface ($\theta_W = 0°$) is not refracted and goes straight. If you look into the water at any other angle, the light ray from the fish to your eye is bent so that the fish is not where it appears to be. The fish is really below where it appears. Similarly, if a person stands in the water—say, up to their hips—and you look at them from outside the water, their legs will appear shorter to you, and the figure is out of proportion.

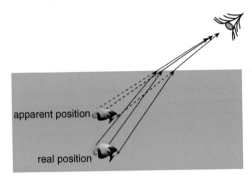

5. The mirror does not invert left to right or upside down, but it only inverts front and back. In the mirror you see the bottom of your face on the bottom and your right ear on the right. Neither up-down nor left-right are reversed. But other people who are *facing you* see your right ear on their left, so they see you differently from the way you see yourself in the mirror.

6. With a prism and white light being refracted and dispersed by it. White light through a prism is refracted away from the apex. In addition, blue light is kinked more than red light, so these, and all other spectral colors, are separated by the prism.

7. Dispersion is the difference between the index of refraction (n) for red and blue light. For example, if n in Plexiglas is 1.55 for blue and 1.51 for red, the dispersion in Plexiglas is $1.55 - 1.51 = 0.04$. Another example: If n in carbon tetrachloride (CCl_4) is 1.38 for blue and 1.32 for red, the dispersion in CCl_4 is 0.06. Note: The *dispersion* in CCl_4 is greater than in Plexiglas, but the index of refraction is smaller. In conclusion, CCl_4 makes a better prism because the difference in dispersion between red and blue is greatest for it, even though it doesn't deflect the entire beam of light as much.

8. *Index of refraction* of a medium, n:

$$n = \frac{velocity\ of\ light\ in\ vacuum}{velocity\ of\ light\ in\ medium}$$

Snell's Law: When light travels from vacuum ($\approx$ air) into a medium,

$$n = \frac{\sin \ of \ angle \ \theta_A}{\sin \ of \ angle \ \theta_M}$$

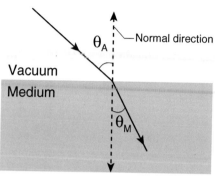

Total internal reflection: When light travels in a medium and encounters an interface with a less dense medium (lower index n), light is refracted out only if the angle θ_M is smaller than a "critical angle"; if θ_M is greater than "critical angle," light is not refracted but only reflected, again in the same medium. That is why this is called total *internal* reflection.

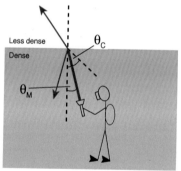

No light enters the less dense medium; this is total internal reflection. The value of critical angle θ_c : $\sin \theta_c = \dfrac{1}{n}$.

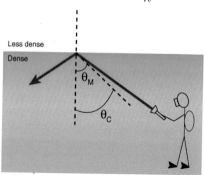

9.

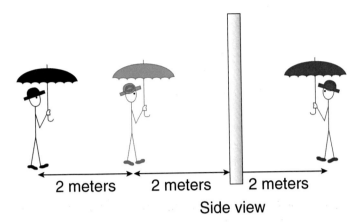

2 meters   2 meters   2 meters

Side view

10.

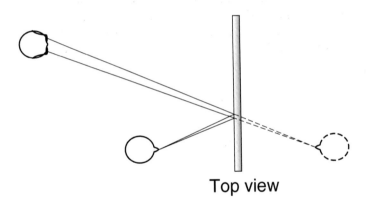

Top view

11.

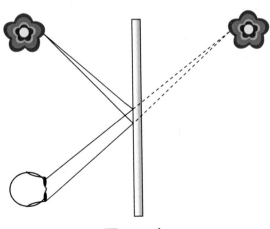

Top view

12.

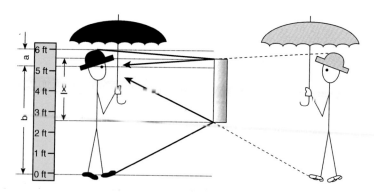

In the preceding diagram,
a = distance from eye to top of hat
b = distance from eye to feet
x = length of mirror

By equal triangles we have $x = \dfrac{a}{2} + \dfrac{b}{2} = \dfrac{a+b}{2} = \dfrac{6ft.}{2} = 3\,ft.$

The mirror must be located so that bottom is 1/2 of the distance from the foot to the eye.

13. In Snell's Law, $\dfrac{\sin\theta_a}{\sin\theta_m} = \dfrac{n_m}{n_a}$ at the critical angle $\theta_m = \theta_c$, $\sin\theta_a = 1$,

therefore, $\dfrac{1}{\sin\theta_c} = \dfrac{n_m}{1}$ and $\theta_c = \sin^{-1}\dfrac{1}{n_m} = \sin^{-1}\dfrac{1}{1.414} = 45°$

14. In Snell's Law, $\dfrac{\sin\theta_a}{\sin\theta_m} = \dfrac{n_m}{n_a}$ at the critical angle $\theta_m = \theta_c$, $\sin\theta_a = 1$,

therefore, $\dfrac{1}{\sin\theta_c} = \dfrac{n_m}{1}$ and $\theta_c = \sin^{-1}\dfrac{1}{n_m} = \sin^{-1}\dfrac{1}{1.52} = 41°$

Chapter 3

1. A lens thicker at the center and thinner at the edges is always converging. A lens thinner at the center is diverging. Therefore, lenses #1, 2, 3, and 6 are converging, and 4 and 5 are diverging.

2. The "strongest lens"—that is, the one with the shortest focal length—is the lens with the smallest radius of curvature and the "curviest," or most convex surfaces. Lens #2 is the shortest focal length.

3. The focal length f is the distance from the center of the lens to the focal point—that is, the point where parallel rays going through the lens converge, for a converging lens, or where rays *seem* to come from for a diverging lens.

See the following diagrams.

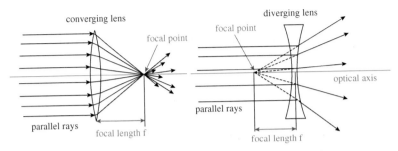

For a converging lens f is always positive; for a diverging lens f is always negative.

4. A strong lens is one that has a very short focal length.

5. Using $\dfrac{1}{i} + \dfrac{1}{o} = \dfrac{1}{f}$ and substituting the values for o and f, we have

$$\frac{1}{i} + \frac{1}{100} = \frac{1}{25}$$
$$therefore \ \frac{1}{i} = \frac{1}{25} - \frac{1}{100}$$
$$and \ \frac{1}{i} = \frac{4}{100} - \frac{1}{100} = \frac{3}{100}$$
$$i = \frac{100}{3} = 33.3 \, cm$$

For the image size we use the magnifcation equation and substitute the values we have for i, o, and h_o:

$$\frac{h_i}{h_o} = \frac{i}{o}$$
$$\frac{h_i}{2 \, cm} = \frac{33.3}{100}$$
$$h_i = \frac{2 \times 33.3}{100} = 0.66 \, cm$$

6. For $o = 100$ cm; $\dfrac{1}{20} = \dfrac{1}{i} + \dfrac{1}{100}$

$\dfrac{1}{i} = \dfrac{1}{20} - \dfrac{1}{100} = \dfrac{5}{100} - \dfrac{1}{100} = -\dfrac{4}{100} = 0.04$

or $i = 25$ cm

7. For $o = 10$ cm; $\dfrac{1}{20} = \dfrac{1}{i} + \dfrac{1}{10}$

$\dfrac{1}{i} = \dfrac{1}{20} - \dfrac{1}{10} = \dfrac{1}{20} - \dfrac{2}{20} = -\dfrac{1}{20}$

or $i = -20$ cm

Note: Negative i means virtual image 20 cm from lens. Image is upright and on the same side of the lens as object.

8. For $f = 5$ cm and $o = 50$ cm we have:
$$\frac{1}{5} = \frac{1}{i} + \frac{1}{50}$$
$$\frac{1}{i} = \frac{1}{5} - \frac{1}{50} = \frac{10}{50} - \frac{1}{50}$$
$$= \frac{9}{50} = 0.18$$

or $i = 5.55$ cm

9. (a)

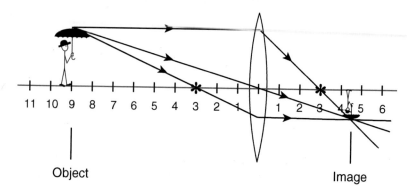

Object   Image

$(f = 3, o = 9, i = ?)$

$$\frac{1}{i} = \frac{1}{3} - \frac{1}{9} = \frac{3}{9} - \frac{1}{9} = \frac{2}{9} \quad i = 4.5 \quad \text{Magification} = \frac{i}{o} = \frac{4.5}{9} = 0.5$$

(b)

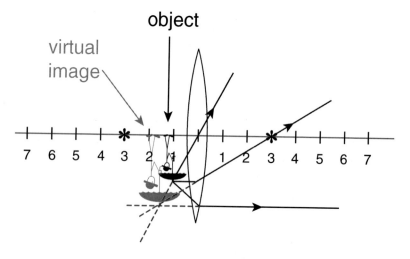

object

virtual image

$(f = 3, o = 1, i = ?)$

$$\frac{1}{3} = \frac{1}{1} + \frac{1}{i}$$

$$\frac{1}{i} = \frac{1}{3} - 1 = \frac{1}{3} - \frac{3}{3} = -\frac{2}{3} \quad i = -1.5 \quad \text{Magnification} = \frac{i}{o} = \frac{1.5}{1} = 1.5$$

Note: We did not use the minus sign to calculate the magnification. We used 1.5, not –1.5. The magnification does not have a sign. It is a "modulus," always between 0 and 1.

10.

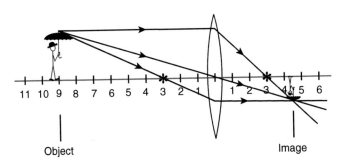

11. This is a little harder: Ray #1 goes through the center of the lens without deviating. Ray #2 strikes the lens parallel to axis and is bent away from axis so that it *appears* to come from the left focal point as shown here.

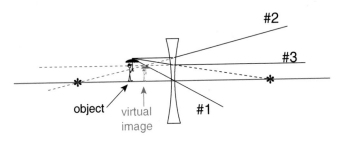

Ray #3 is a little trickier: The ray that is aimed toward the right focal point (black, dashed line) is bent so that it leaves the lens parallel to the axis! To see this think of ray #2 going in the reverse direction. Now extend the three rays back (magenta, dashed lines) to their meeting point to find the image position and size. The image is upright, smaller, closer to the lens, and virtual.

12.

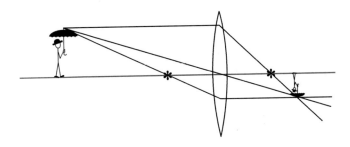

13. Use similar triangles:

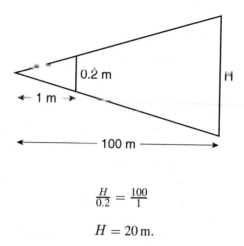

$$\frac{H}{0.2} = \frac{100}{1}$$

$$H = 20\,\text{m}.$$

The oak tree is 20 m high.

Chapter 5

1. "Film speed" refers to the amount of light required to produce a certain amount of blackening of the film; high-speed or fast film is a film that requires little light to darken, or a very short exposure time. That is, fast film has greater sensitivity to light than slow film. Fast film has disadvantages: it is grainy and never deep black because the emulsion is thin.

2. The exposure time is the amount of time that the shutter stays open—in other words, the time that the film is exposed to light, to the image.

3. $f/ = f/D$, where f is the focal length of the lens and D the diameter of the iris. A large f number means a small diameter and vice versa. f numbers vary following the sequence: $f/2$ $f/2.8$ $f/4$ $f/5.6$ $f/8$ $f/11$ $f/16$ $f/22$ $f/32$.

   Changing the aperture by two steps means changing the diameter by a factor of 2. Therefore, changing the aperture by one step means changing the diameter by a factor of 1.4, and the area, proportional to the square of the diameter, by a factor of $1.4^2 = 2$. The lens area changing by a factor of 2 corresponds to the illumination varying by a factor of 2 (→amount of light entering per second). For example, changing the f number from $f/2.8$ to $f/11$, there are four steps and $f/11$ is $2 \times 2 \times 2 \times 2$ times less light than $f/2.8$, or 16 times less light.

4. Depth of field is the range of object distances that produce adequately focused images on the film.

5. Changing the exposure time from $1/60 \rightarrow 1/250$ corresponds to having two steps less light on the film. Therefore, the aperture must also be changed by two steps in order to compensate for this, from $f/8$ to $f/4$.

6. Exposure time from $1/250 \rightarrow 1/30$ corresponds to 3 steps more light on the film. The f number must be changed from $f/1.4$ to $f/4$.

7. The ISO 400 film is "faster"; it requires four times less light. It has change by two steps. So we must change the f number by two steps toward lower illumination: from $f/11$ to $f/22$.

8. ISO $200 \rightarrow 400$ is twice as sensitive. This is equivalent to having one step more light, so the f number must be changed from $f/11$ to $f/16$, corresponding to one step less light.

9. Picture (a) has the same exposure as (b). The difference is three steps in exposure time: $60 \rightarrow 125 \rightarrow 250 \rightarrow 500$ and three steps in f number $16 \rightarrow 11 \rightarrow 8 \rightarrow 5.6$. However, fast moving cars will be blurred, because 1/60 sec is too long an exposure time. Picture (c) is underexposed by one step (lower film illumination): Shorter exposure time is not compensated by a change in f number. This picture will be darker than pictures (a) and (b).

10. Larger f number gives larger depth of field, so your friend and the landscape behind him are both in focus if you use exposure (a).

11. ISO $100 \rightarrow$ ISO 800 requires three steps less light.
$f/2 \rightarrow f/4 = 2$ steps less light
Therefore the exposure should be one step shorter $1/30 \rightarrow 1/60$ sec.

12. (a) $f = 5\,\text{cm}$ and $i = 3\,\text{m} = 300\,\text{cm}$. Then $1/o = 1/5 - 1/300 = 0.197; o = 5.08\,\text{cm}$. The magnification is $i/o = 59$. So if the slide is 35 mm by 24 mm, the image will be 59 times that size: $35 \times 59 = 2065$ mm wide by $24 \times 59 = 1416$ mm high.
(b) The aperture diameter is the focal length divided by the f number. Here the diameter is $(5/2)\,\text{cm} = 2.5\,\text{cm}$. The $f/2$ lens transmits more light than and $f/4$ lens, but $f/2$ is not necessarily better; larger apertures have worse depth of field and allow more aberration and distortion.

13. The exposure time was *decreased* by one step to 1/250 s. To compensate, we should *increase* the aperture size by one step: $f/16$ goes to $f/11$.

14. The film speed goes from ISO 100 to ISO 400. Because the second film is more sensitive this is equivalent to having 2 steps more light. The

exposure time decreases, thus providing 1 step less light, so the aperture diameter must decrease by one step, and the $f/$ increase by 1 step. $f/11$ goes to $f/16$.

15. The film speed goes from ISO 200 to ISO 400, so the film needs 1/2 as much light: one step less.
The exposure time *increases* by one step, so the aperture size must *decrease* by two steps: $f/11$ goes to $f/22$.

Chapter 6

1. 400 to 700 nanometers (abbreviated nm). The metric prefix *nano* means multiply by 10^{-9}. $1\,\text{nm} = 10^{-9}\,\text{m}$—that is, a billionth of a meter.
2. Red: 600–700 nm, Green: 600–500 nm, Blue: 400–500 nm.
3. The eye color sensors are called cone cells.
4. See Figure 6.4.
5. The type II cone cells (incorrectly called "green sensors") are sensitive all the way from 400 to 660 nm, including violet, indigo, blue, green, yellow, and orange. They are, however, most sensitive around 550 nm, which is green.
6. Lights of two complementary colors added together give white light. This works for lights, for additive color mixing! Not for adding paints! The complementary colors are R and C, G and M, and B and Y.

Chapter 7

1. Magenta. $R + B = M$.
2. Say you selected orange: On the color triangle you see that orange is between Y and R. Orange $= Y + R$, but $Y = R + G$, so orange $= 2$ parts red $+ 1$ part green. In 8-bit graphics, $R = 255, G = 128, B = 0$.
Magenta: $R = 255, G = 0, B = 255$.
Blue: $R = 0, G = 0, B = 255$.
Green: $R = 0, G = 255, B = 0$.
Cyan: $R = 0, G = 255, B = 255$.
Pale yellow: $R = 255, G = 255, B = 150$.
Pale pink: $R = 255, G = 200, B = 200$.

3.

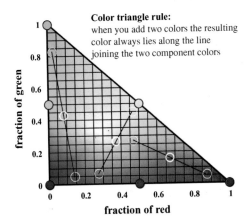

Color triangle rule:
when you add two colors the resulting
color always lies along the line
joining the two component colors

fraction of green (y-axis)

fraction of red (x-axis)

4. The complementary color is located directly across from white. The colors are as follows: A′ is green of high saturation; B′ is bluish cyan of high saturation; C′ is blue of high saturation; D′ is greenish cyan of low saturation.

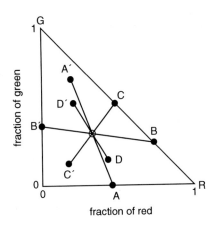

fraction of green (y-axis)

fraction of red (x-axis)

5. (a) r = 0.3, g = 0.1. This is bluish magenta of high saturation (see color triangle on the right)
  (b) r = g = b = 0.33. This is white (zero saturation).
  (c) r = 0.1, g = 0.5. This is greenish cyan of high saturation.
  (d) r = g = 0.4. This is yellow of low saturation.

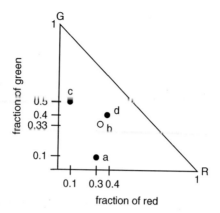

6. You could only mix colors within the gray triangle below. These include some of the purple colors, white and orange, but not cyan.

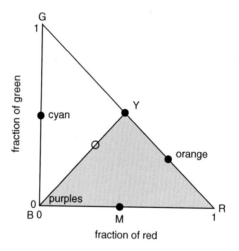

Chapter 8

1. A rapid way to identify the color of the light is to simplify the spectrum and say that 400–500 nm is blue, 500–600 nm is green, and 600–700 nm is red. Filter A transmits half the light in the green region, and all the light in the red region. This corresponds to a fraction of green $g = 1/3$, and a fraction of red $r = 2/3$ (twice more red than green). This is orange on the color triangle.

    Filter B transmits all the light in the green region, 500–600 nm, and is therefore green.

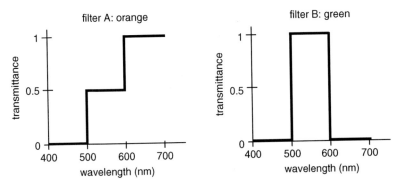

2. Filters A and B are orange and green, as determined in problem 1. To find the combined transmittance, we *multiply* the transmittances of the two filters in each wavelength region. In the B region (400–500 nm) they are both zero, so the combined transmittance is zero. In the G region (500–600 nm) filter A's transmittance is 1/2 and filter B's transmittance is 1. The combined transmittance is 1/2. In the R region (600–700 nm) the filter A's transmittance is 1 and filter B's is zero: the combined transmittance is zero. The final transmittance curve looks like the one on the right.
The combination is dark green.

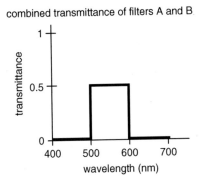

combined transmittance of filters A and B

Another way to look at this: Filter B absorbs all the light except for green, and filter A absorbs half the green light. Combining the two filters, half of the green light goes through and none of the rest (R and B).

3. Now we add the lights from the two filters. A transmits 0.5 units of green and 1 unit of red. B transmits 1 unit of green. The total light on the screen, then, is (0.5 + 1 = 1.5) units of green and (1 + 0 = 1) unit of

red. To name the color, we calculate the fractions of red and green, and
find the color on the color triangle. The total amount of light is (1.5 +
1 = 2.5) units, so the fraction of red is (1/2.5 = 0.4) and the fraction of
green is (1.5/2.5 = 0.6). We see on the color triangle that this is greenish
yellow of high saturation.

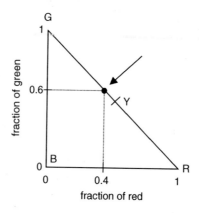

4. Filter C transmits all the light in the blue and green regions. Since it
transmits equal amounts of green and blue, it is a cyan filter.
Filter D transmits all the green and the red light, in equal amounts, and
is therefore yellow.
Subtractive mixing: Filter C absorbs all the red light and filter D
absorbs all the blue light, and neither absorbs green light. The combined
transmittance curve looks like the following:

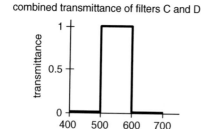

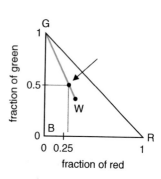

Additive mixing: Filter C transmits 1 unit of blue and 1 unit of green,
Filter D transmits 1 unit of green and 1 unit of red. In total there are 1 unit
of blue, 2 units of green, and 1 unit of red light (total 4 units). The fraction of green is g = 2/4 = 0.5 and the fraction of red is r = 1/4 = 0.25.
On the color triangle above we see that this is green of low saturation.

We could also see this because one unit of each primary, blue, green, and red, goes to form white, and 1 unit of green is left over. This color is therefore G + W.

5. Filter E is blue. It transmits light from 400 to 500 nm. Filter F is red. It transmits all light between 600 and 700 nm.

   Subtractive mixing: Filter E absorbs all light above 500 nm, and Filter F absorbs all light below 600 nm. If you overlap them, no light can go through.

   Additive mixing: Filter E transmits 1 unit of blue light, and filter F transmits 1 unit of red light. The additive combination is 1/2 blue and 1/2 red, which is magenta.

6. The steps in the transmittance curves no longer come right at the 400–500–600 nm boundaries, but we can still see how much light is transmitted in each range. Filter H transmits 1/2 the light in the green region and 1/2 the light in the red region. It is therefore dark yellow. Filter G transmits all the light between $\sim$570 and $\sim$640 nm, no light between 640 and 680 nm, and all the light above 680 nm. This filter is also yellow, although its transmittance curve looks very different! This color is reddish yellow, since it has more red than green wavelengths. It appears yellow-orange, like an egg yolk.

   Subtractive mixing: Filter H absorbs all the light below 500 nm and half the light between 500 and 700 nm. Therefore, we multiply the transmittance of filter G by 0.5, and the transmittance curve looks like this:

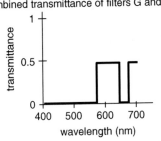

combined transmittance of filters G and H

This is brown, or darker egg yolk yellow.

Additive mixing: Filter G transmits 1 unit of green and 1 unit of red, Filter H transmits 1/2 unit of green and 1/2 unit of red (total 3 units). The resulting light on the screen will have a fraction of red $r = 1.5/3 = 0.5$ and a fraction of green $g = 1.5/3 = 0.5$. This is therefore yellow.

7. You added 2 parts of C and one of Y paint. Think subtractively! Starting from the line indicating white light (illumination), go down, following the red arrows, by 2 parts of C and 1 part of Y. You have

therefore removed all red light (absorbed by C paint), half of the B light, and none of the G light. Remember that YMC are complementary of BGR, and each color absorbs its complementary color.

The resulting reflectance curve for this paint mix is the below second

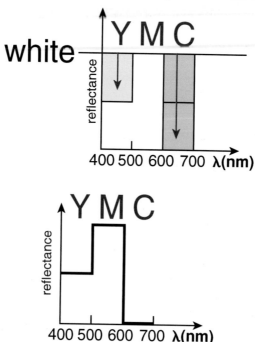

Now, to understand what color this is, think additively again, and read the reflectance curve from the bottom, as shown on the right. Your eye is receiving 1 part of B light, 2 parts of G light, and 0 parts of R light. It is now easy to find this color on the color triangle and interpret it as cyan-green, or aqua.

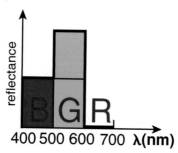

Let's look at it numerically: $S = 1B + 2G + 0R$

The fractions of red and green, therefore, are $r = 0$ and $g = 2/3 = 0.66$.

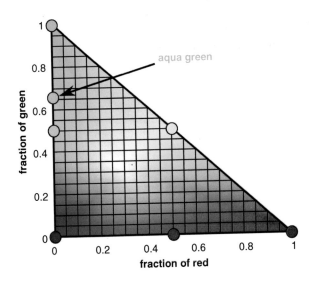

This color is between C and G. The reflectance curve above may have already suggested to the careful reader that this color is C + G. There are 2 parts of C light (B + G) and 1 additional part of G light diffusely reflected by this paint mix. This is why the resulting color on the triangle is closer to C than to G.

This is a trick that always works: a reflectance curve can always be interpreted as the *sum of lights* in the B, G, and R regions, and therefore be treated according to additive color mixing. In conclusion, we used subtractive color mixing to determine the resulting reflectance curve for a paint mixture but then determined what color that curve represents by using additive color mixing: the color triangle and the BGR light components. This is because the human eye and brain perceive and interpret color additively.

Chapter 10

1. 250 Hz, 0.25 kHz; 7.7×10^{-4} ms $= 0.77 \ \mu s$
2. 40 kHz; 2.5×10^{-5} s, 0.025 ms, 25 μs.
3. 0.01s, 10 ms; 100 Hz, 0.1 kHz, 8 units.
4. 2000 m, 2.5 cm, 7.8 kHz; 0.015 m, 2.6 kg, 0.3 ms.
5. 50 ms; 50 μs.
6. 40 oscillations; 200 oscillations.

Chapter 11

1. $k = \dfrac{F}{x} = \dfrac{2N}{0.06\,\text{m}} = 33\dfrac{N}{m}; f = \dfrac{1}{2\pi}\sqrt{\dfrac{k}{M}} = 1.3\,\text{Hz}.$

2. $f = 50\,\text{Hz}; f = 31.6\,\text{Hz};$ larger; higher

3. 686 N; 113.4 N.

4. $k = \dfrac{F}{x} = \dfrac{2N}{5\,\text{cm}} = 0.4\dfrac{N}{\text{cm}}; \dfrac{2N}{0.05\,\text{m}} = 40\dfrac{N}{m}.$

5. $2.0\,\text{Hz}; \dfrac{2.0}{\sqrt{2}}\,\text{Hz} = 1.41\,\text{Hz}; 2.0 \times \sqrt{2}\,\text{Hz} = 2.82\,\text{Hz}.$

Chapter 12

1. (a) 1/4, 1/16; (b) same: pitch of tone does not change as tone dies away.

2. 520 oscillations: 1 mm, 0.5 mm, 0.125 mm.

3. Frequency of driving force; amplitude of oscillation; 60 Hz; 20 Hz;
$\tau = \dfrac{0.4}{\Delta f} = \dfrac{0.4}{20} = 0.02\,s = 20\,\text{ms}.$

Chapter 13

1. 1000 Hz; 2000 Hz; 1000 Hz.

2. Simple number ratios like 2:3 cause no beats and thus are perceived as harmonious.

3. 444 Hz or 436 Hz.

4. Shift is 2 divisions on horizontal axis, while full oscillation is 12 divisions. $\dfrac{360°}{12} \times 2 = 60°.$

Chapter 14

1. $v = 15\,\text{cm/s}.$

2. Flute: longitudinal; string: transverse.

3. $\dfrac{344\,\text{m/s}}{20\,1/_s} = 17.2\,\text{m}; \dfrac{344\,\text{m/s}}{20000\,1/_s} = 0.017\,\text{m} = 1.7\,\text{cm}.$
$f = \dfrac{344}{0.75}\,1/_s = 459\,\text{Hz}$

4. 4x greater.

5. (a) $\rho = \frac{M}{\ell} = \frac{0.05\,\text{kg}}{2\,\text{m}} = 0.025\ \text{kg}/\text{m}$

    (b) $v = \sqrt{\frac{F}{\rho}} = \sqrt{\frac{300N}{0.025\ \text{kg}/\text{m}}} = 109.5\ \text{m}/\text{s}$

6. (a) $\rho = \frac{M}{\ell} = \frac{2\,\text{kg}}{500\,\text{m}} = 0.004\ \text{kg}/\text{m}$

    (b) $\sqrt{\frac{40}{0.004}} = 100\ \text{m}/\text{s}$

    (c) $2^2 = 4x$ higher tension $= 160$ N.

7. Loud if $6.0\,\text{m} - 5.4\,\text{m} = 0.6\,\text{m} = n\lambda$ with $n = 1, 2, 3, \dots.$

    $\lambda = \dfrac{0.6\,\text{m}}{n} = 0.6\,\text{m}; 0.3\,\text{m}; 0.2\,\text{m};$ etc.

    $f = \dfrac{344}{\lambda} = 573\,\text{Hz}; 1147\,\text{Hz}; 1720\,\text{Hz};$ etc.

    Soft if $6.0\,\text{m} - 5.4\,\text{m} = \frac{1}{2}\lambda; \frac{3}{2}\lambda; \frac{5}{2}\lambda;$

    $\lambda = 1.2\,\text{m}; 0.4\,\text{m}; 0.24\,\text{m};$

    $f = \dfrac{344}{\lambda} = 287\,\text{Hz}; 860\,\text{Hz}; 1433\,\text{Hz};$

Chapter 17

1. $\dfrac{240\,\text{Hz}}{300\,\text{Hz}} \times 60\,\text{cm} = 48\,\text{cm}; \dfrac{60}{40} \times 240\,\text{Hz} = 360\,\text{Hz};$

    $\dfrac{60}{20} \times 240 = 720\,\text{Hz}.$

2. $\sqrt{4} = 2$ times lower $f = 100\,\text{Hz}$; half the frequency; $200\,\text{Hz} \times \sqrt{\dfrac{300}{200}} =$

    $245\,\text{Hz}$; four times more mass $\rightarrow$ half the frequency $= 50\,\text{Hz}.$

3. $\dfrac{240\,\text{Hz}}{\sqrt{2}} = 170\,\text{Hz}.$

4. 4th, 8th, 12th missing, 2nd strongest.

5. 30 cm; 40 cm; 45 cm measured from bridge.

6. Overwrapped (or overwound) string; stiffness raises frequency; not desirable.

7. 2nd; 6th; 10th.

Chapter 18

1. (a) 215 Hz; 172 Hz.

    (b) 107.5 Hz.

(c) The frequency remains almost the same. If one takes into account the correction mentioned in Ch. 18.10, the answer is 213.4 Hz and 211.8 Hz for the 2 cm and the 4 cm diameter pipe, respectively.

(d) Increases frequency.

2. (a) 260.6 Hz

   (b) Closed pipe

3. (a) 0.86 m.

   (b) 100 Hz.

4. One can overblow to higher modes: $2f, 3f, 4f$, etc.

5. 4.3 m if it is an open pipe, 2.15 m if it is closed.

6. (a) 286.7 Hz, 573.3 Hz, 860 Hz.

   (b) 143.3 Hz, 430 Hz, 716.7 Hz.

7. (a)

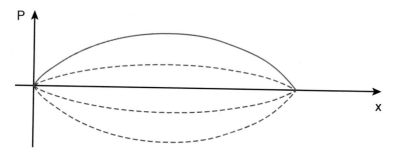

(b)

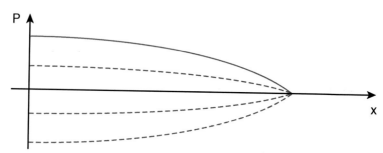

8. Speed of sound 344 m/s at 20°C, 354 m/s at 37°C. Frequency increases in proportion:

$$f = \frac{354}{344} \times 260 \, \text{Hz} = 267.6 \, \text{Hz}.$$

Chapter 19

1. As the superposition (sum) of sinusoidal oscillations, whose frequencies are multiples of the fundamental frequency $f = \frac{1}{T}$. In this example, for $T = 5$ ms, $f = 200$ Hz, 400 Hz, 600 Hz, etc.

2.

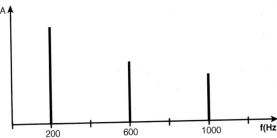

3.

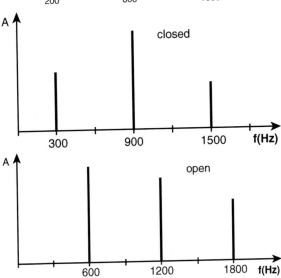

4. 300 Hz, 600 Hz, 900 Hz, etc.

5.

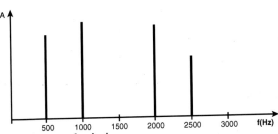

i.e. 3x, 6x fundamental missing

Chapter 20

1. (a) $D_5 = \frac{9/4}{5/3} \times 440\,\text{Hz} = \frac{9 \times 3}{5 \times 4} \times 440\,\text{Hz} = \frac{27}{20} \times 440\,\text{Hz} = 594\,\text{Hz}$.

    (b) $F_4 = \frac{4/3}{5/3} \times 440\,\text{Hz} = \frac{4}{5} \times 440\,\text{Hz} = 352\,\text{Hz}$.

    (c) $f = \frac{6}{5} \times \frac{440\,\text{Hz}}{4} = 132\,\text{Hz}$.

2. 30 cm from bridge; $\dfrac{60\,\text{cm}}{(1.0595)^2} = 53.4\,\text{cm}$ from bridge.

3. 400 Hz, 600 Hz, 800 Hz, 1000 Hz.

4. G_4, D_5, G_5, B_5.

5. $\dfrac{220\,\text{Hz}}{(1.0595)^2} = 196\,\text{Hz}$, $\ell = 344(\text{m/s})/(2 \times 196\ \text{Hz}) = 0.878$ m

6. (a) $(350 \times 1.0595 - 350)\,\text{cm} = 20.8\,\text{cm}$
    (b) 42.9 cm
    (c) 66.3 cm

7. 12; 7; 5; 4; 3.

Chapter 21

1. Fourth, fourth, fourth, major third, fourth.

2. Just tuning: no. $C - D = 9/8$; $D - E = 10/9$.
    Tempered: yes.

3. 5 semitones $\dfrac{60\,\text{cm}}{(1.0595)^5} = 44.9\,\text{cm}$ from bridge. It is A_4.

4. (a) Length ratio $80/60 = 4/3$;
    (b) Fourth above $G_3 = C_4$;
    (c) Octave above $C_4 = C_5$.

5. $(150 \times 1.0595 - 150)\,\text{cm} = 8.93$ cm.

6. Assuming the musician "buzzes" the lips near the right frequency, the pressure wave reflected from the instrument locks the lip frequency to the pipe frequency.

Index

Made in the USA
Lexington, KY
25 January 2018